PRAISE FOR GWENDOLYN BOUNDS AND
Little Chapel on the River

"A compelling post-9/11 story. . . . This is a sweet tale of how one woman reorganized her personal life in the midst of national upheaval."   —*Chicago Tribune*

"This is a stirring . . . story of lives that might otherwise have gone on, unnoticed, had Bounds not happened in for that quick beer. We are lucky she did."
—*New Orleans Times-Picayune*

"Bounds's elegiac tale of transformation is a story filled with sweet surprises that never becomes cloying."   —*New York Post*

"Timely and meaningful. . . . Bounds sketches the pub's regulars with humorous, compassionate strokes, and questions—in light of this place so slow to change and stubborn in its values—whether the fast track to widespread homogenization is really the route we should be traveling."   —*Library Journal*

"A true romance—with a place."   —Associated Press

"In an age of spiky-heeled chick-lit, this book is a refreshing change."
—*Milwaukee Journal Sentinel*

"Bounds's tribute to this community love affair speaks to the deep-seated craving all people have for a special place as well as its boundaries."   —*Pittsburgh Post-Gazette*

"Bounds's story . . . modestly reminds us that in this uncertain world, when you come to a place that speaks to you, you should hold it dear and treasure it while it lasts."   —*Publishers Weekly*

"Over beers and smokes . . . life stories bounce around the bar with the mock-insults of people who've known one another more than the forty years Guinan's [has] been in business. . . . Without gauzy romanticism, Bounds captures the warmth of the place and the rootedness it symbolizes."   —*Booklist*

"A tender, heartfelt love letter to a small bar in the small town of Garrison, New York. . . . Absolutely gripping. . . . The book's gentle appeal will charm."
—*The Toronto Star*

P9-COO-858

"*Little Chapel on the River* is a layered and vivid chronicle of perseverance and possibility. With a reporter's eye for detail and a woman's open heart, Bounds weaves the strands of disparate lives, her own included, into a seamless, shining tale."
—Nancy Cobb, author of *In Lieu of Flowers: A Conversation for the Living*

"Stunning. *Little Chapel on the River* is beautifully written, artfully crafted, and lovingly told—a sweet, caring infectious hymn about a place and people most of us would be too preoccupied to notice in the rush to grab a newspaper and make it to a train, to get someplace else. Luckily—for her and us—Gwendolyn Bounds stopped for a beer, and a story that began in flight and fear ends in sanctuary and hope. This is the rare first-person book that forces you to take a deep breath and reconsider how you live your life."
—Stefan Fatsis, author, *Word Freak: Heartbreak, Triumph, Genius, and Obsession in the World of Competitive Scrabble Players*

"Every writer knows there is no more compelling story than an honest and close observation of a life, and Gwendolyn Bounds has surely done this with her account of the Guinan family as they provide a rare glue that binds a small group of very decent people in a small Hudson River town. *Little Chapel on the River* is a book as natural and as consequential as the trees and the railroad tracks that border the place itself, charged with writing that glows with sensitivity and rings with truth. And as American as July Fourth, it is a gem of a story too, one that you will read in one day as I did, for I could not put it down."
—Dennis Smith, author, *Report from Engine Co. 82, A Song for Mary,* and *Report from Ground Zero*

"Reading Gwendolyn Bounds's very fine book is much like a delightful night spent visiting a pub in Ireland: You remember it as you remember time spent in good company out in God's country."
—Frank Gannon, author, *Mid-Life Irish: Discovering My Family and Myself*

"Gwendolyn Bounds has perfectly captured the sounds, flavors—indeed, the soul—of a quickly disappearing kind of small town life."
—Billy Collins, poet laureate, author of *Picnic, Lightning*

"Set aside a huge chunk of time to read *Little Chapel on the River* as it is a so superbly woven tale that putting it down would cause heartache."
—Malachy McCourt, author of *A Monk Swimming*

Michael F. Bounds

The Chapel's longtime owner, Jim Guinan, seated at his bar with the author

About the Author

GWENDOLYN BOUNDS is a columnist for the *Wall Street Journal,* where she has worked since 1993. A native of North Carolina, she now lives in New York's Hudson River Valley.

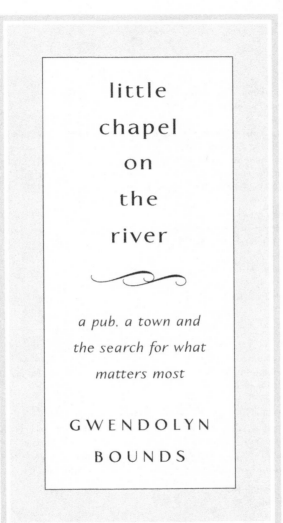

little chapel on the river

*a pub, a town and
the search for what
matters most*

GWENDOLYN
BOUNDS

HARPER

NEW YORK • LONDON • TORONTO • SYDNEY

HARPER

A hardcover edition of this book was published in 2005 by William Morrow, an imprint of HarperCollins Publishers.

LITTLE CHAPEL ON THE RIVER. Copyright © 2005 by Gwendolyn Bounds. All rights reserved. Printed in the United States of America. No part of this book may be used or reproduced in any manner whatsoever without written permission except in the case of brief quotations embodied in critical articles and reviews. For information address HarperCollins Publishers, 10 East 53rd Street, New York, NY 10022.

HarperCollins books may be purchased for educational, business, or sales promotional use. For information please write: Special Markets Department, HarperCollins Publishers, 10 East 53rd Street, New York, NY 10022.

FIRST HARPER PAPERBACK PUBLISHED 2006.

Designed by Jo Anne Metsch

The Library of Congress has catalogued the hardcover edition as follows:

Bounds, Gwendolyn.
    Little chapel on the river : a pub, a town and the search for what matters most / by Gwendolyn Bounds.—1st ed.
        p.   cm.
    ISBN 0-06-056406-7 (alk. paper)
    1. Garrison (N.Y.)—Social life and customs.   2. Bars (Drinking establishments)—New York (State)—Garrison.   3. Irish Americans—New York (State)—Garrison—Biography.   4. Garrison (N.Y.)—Biography.   5. Bounds, Wendy—Travel—New York (State)—Garrison.   6. Garrison (N.Y.)—Description and travel.   I. Title.

F129.G1855B68 2005
974.7'32—dc22

                                                                                          2004062812

ISBN-10: 0-06-056407-5 (pbk.)
ISBN-13: 978-0-06-056407-0 (pbk.)

06   07   08   09   10   ❖/RRD   10   9   8   7   6   5   4   3   2   1

In memory of my grandfather
William Albert McKnight
and my godmother and namesake,
Gwendolyn Jenkins Duffey.
You would have loved this place.

contents

part II

part III

author's note

While Guinan's square footage may be small, the legion of lives touched inside its walls is not. This little book chronicles a pivotal moment in time for the place and the family who runs it, as well as for the author. Many volumes would be required to capture all the tales settled in the old pub's corners. Because of that, this one story asks that the experiences of a few be allowed to represent the whole, and the author acknowledges first and foremost all those missing in these pages. Deep thanks to everyone who shared their memories with me, but also to those I've never met who have purchased papers or sipped beers there through the years. Your loyalty helps the place's heart to keep beating. Sincere admiration also goes to the many Irish Night musicians who have trekked from all corners of this region to play in the great hall of Guinan's—you are worthy of a book of your own.

To anyone who has ever known a spot like this, a spot that makes you feel more at home sometimes than home itself, I'd just like to add, go there if you still can. Be there. And don't wait for tomorrow. Go today.

little chapel on the river

the bar

. . . this tastes like piss-water.

Hey, is this a Ballantine?" the hairy guy shouts, waving the green bottle in my face.

Sweat begins to gather in small beads along my lower back. I dig frantically in the red metal cooler, my slender hands trying to recall the chilly terrain. *Harp in left back. Budweiser on the right, Guinness cans beside the Murphy's Irish Stout. Rolling Rock mixed in somewhere with the Ballantine.*

"I wanted a Rolling Rock," he says, leaning over the bar, stubbly face in mine, breath heavy and sweet from a couple of hours of drinking. "I told you I wanted a ROLLING ROCK."

THE TINY PUB is crowded, and it's only 6:15 P.M. A light breeze sneaks through the windows off the Hudson River. On the opposite bank, I can see the fortresses of the West Point military academy glowing hot in the late summer sun. The yellow rays streak across the water and onto the backs of the men standing in this green-walled, green-ceilinged Irish drinking hole nestled between the river and railroad tracks. The pub, which is barely big enough to hold the old giant metal Coca-Cola cooler stuffed with beer, red-topped bar and five stools, is tacked onto the side of an old country store—almost as an afterthought. The seventy-six-year-old owner lives upstairs, as he has for more than four decades.

Today's crowd is Friday's typical motley mix of local blue-collar guys, boaters from the makeshift yacht club out back and commuters just off the train from New York City an hour south. They stand close but not touching, their collective posture one of possession and fine-tuned ease.

I know most of them by this point. There's Fitz, a tough-talking former U.S. federal marshal whose body is a topological map of scars. Most visible is the bite wound on his right forearm from when he broke up a fight between his unneutered dogs, Buck and Ranger. It's impressive, though not as thick as the scar around his knee where AK-47 fire hit unexpectedly in Vietnam. There are others, scars from Vietnam that is, but this last one is what he might show you when the beer is going down good and the memories anesthetized.

Two stools down from Fitz sits Dan, the white-haired liberal lawyer who drinks ten-ounce Pepsis and eats serial packages of Fig Newtons. Dan is more of a scotch and wine man himself but comes to this beer bar to escape his depositions and trade barbs with the ultraconservative Fitz. Currently, the lawyer is chatting with Ed, one of the two handsome Preusser brothers, whose mom runs the oldest high-end real estate agency in town. And at my right elbow I see, or rather sense, Old Mike, a hearty, good-natured fellow who always stands sentinel at the bar's end, Schaefer bottle in hand, like a traffic cop. He was introduced to me as Old Mike because there used to be a Young Mike—but no-body talks about that much inside here.

The one person missing is Jane, the bar's regular busty bar-tender, who's running late today, which is why I'm standing back here screwing up orders.

The truth is, I've never tended bar in my life until this moment. In fact, since I got out of college nine years ago, I haven't done much of anything except churn out stories for the same national newspaper about the wheelings and dealings of big corporations. Right now, suffice it to say I always thought bartending would be a little easier. After all, it's just bottled beer at this joint. Pop the caps, take the money, smile a little. Right?

Instead, I'm frantic, trying to remember five orders at once, carry on multiple conversations and quickly add strings of $3.25s and $2.75s in my head each time someone buys a round. But rule number one here: no calculators. So I keep ticking off numbers on my fingers. And I'm getting confused because the guys have bought me a couple of beers, and they're going to my head. I don't know where Jane is. And now I've just served the wrong brand to some grizzly boater with the hairiest arms I've ever seen.

I LOOK MISERABLY at the open Ballantine bottle in his hand.

They're both green, I offer lamely, finally laying my hands on a Rolling Rock and hoping he'll be a good sport and cut me a break.

No chance.

"Yeah, except this tastes like piss-water," he says, plunking the Ballantine down on the counter before me.

My face reddens as the other guys laugh. They watch for a moment, waiting, and I sense I'm flunking some unstated, crucial test to hold my own back here.

Fine, I say lightly, setting the Ballantine aside and hoping no one notices my hand shaking. I'm just wondering, though, I ask, forcing myself to meet his bloodshot eyes. How do you know what piss-water tastes like?

A pause. And then the tide turns. "Heh *heh heh.*" Fitz's trademark laugh sends my nemesis retreating to the back of the bar with his Rolling Rock. The veteran keeps peace, though, by buying everyone a round, including the hairy stranger. Ed Preusser checks his watch and says he'll stay for just one more. So I gratefully begin pulling out beers again, stacking coasters to keep count of each person's beer order.

Another rule here: no written bar tabs allowed.

WHEN I FINALLY look up again, I see the white-haired proprietor moving slowly in his kitchen, which opens into the rear of the pub. He takes the silver teakettle off the stove and pours hot water into his cup. Then he pokes his head into his bar, checks out the clientele and calls hello to a few. They all answer like respectful schoolchildren: "Hi, Jimmy . . . Hello, Jim . . . How ya feeling, Guinan?" Jim catches my eye. He nods and shouts over their heads—"You're doing just fine, luv"—and retreats back into his living room. After he goes, I slip a few dollars from my pocket into the wooden box that passes as the bar's cash register. That's to cover the Ballantine mix-up. Then I notice Fitz's Beck's is nearly empty. Leaning across the bar, I put my hand gently on his arm and invoke one more unwritten rule.

"Next one's on the house," I say.

THIS IS THE story of a place, the kind of joint you don't find around much anymore, a spot where people wander in once and return for a lifetime.

For most of its days, the place billed itself as a country store, but its true heart was the adjacent pub. There was a rusty horseshoe posted above one door and a gold shamrock embedded, slightly off center, in the fireplace hearth. The floor slanted toward the river, and the men returned to the same seats every Friday. Most people called it Guinan's (sounds like Guy-nans) after the Irish owner, Jim Guinan. Some called it the bar. One regular patron christened it his "riverside chapel," which seemed to me to fit best because for most of these guys, coming to Guinan's *was* something of a religion, with its own customs, community and rites of passage. There was even a pastor of sorts—Jim—who on a good night could tell a story that might run as long as a Sunday sermon.

Folks had been congregating at Guinan's quite a while before I showed up—forty-two years, to be precise—which was long enough for the place to have a memory and a cast of characters as constant as the hourly trains rumbling by its windows. For them, the cramped space was far more than a pit stop on the way home—it was an extension of home itself. Guinan's was where they came after a death to toast and remember, on holidays and birthdays to pay their respects and buy a round or two, or on a late winter afternoon when a cold wind made things lonely enough that you just needed to see a friendly face. When she was alive, Jim's wife, Peg, would welcome the men, scold them if their language turned rough and offer supper to those who had nowhere else to be. Here inside this family's mismatched stucco green walls, it was always safe.

When I stumbled upon this world in late 2001, I didn't know that all of this was on the brink of disappearing. And I wasn't looking for a story. In fact, all I really wanted was a quick beer and to get back to New York City. What came next—upending my life because of this hole-in-the-wall pub—suffice it to say, was never supposed to happen. At least not by any plan I'd laid out.

But I'm getting ahead of myself.

There is a beginning, so let's start there. It is morning, and *the sky is a brilliant blue and clear, the air unusually warm for September. . . .*

part

I

On the Hudson

there was always

the opportunity to be

educated deeply in the heart.

—MARK HELPRIN,
Winter's Tale

1

home

Didn't you hear it?

The sky is a brilliant blue and clear, the air unusually warm for September—a sign we are still closer to August than October. A little past 8 A.M. finds me still moving slowly around my Manhattan apartment, stepping over the dirty clothes and half-unpacked suitcases from a two-week beach vacation in the Hamptons and Southern California. My girlfriend, Kathryn, is in the kitchen washing dishes. Neither of us hurry. The *Wall Street Journal*'s offices, where we work, sit directly across the street, making our commute something approximating eight minutes from door to desk.

It is Tuesday.

Cup of coffee in hand, I curl up in a chair at our dining room table overlooking the harbor in Battery Park City. To my left is the Hudson River. To my right, the World Trade Center towers. The windows are open to catch the early morning breeze. Joggers run by, breathing in damp sea air. Workers are streaming off the ferry, headed into the various office buildings downtown.

Sipping my coffee, I glance at the *New York Times* front page for September 11, 2001. A collage of pictures shows the city's mayoral candidates stumping for last-minute votes for the day's primary elections. There is a story about stem cell research, and a piece about the trafficking of nuclear material to Iraq and Iran. I skim an article about young girls dressing like Britney Spears at school and think about my day, which spreads before me, orderly and full. There's a 9:30 A.M. doctor's appointment, a 1 P.M. lunch at Odeon, editing in the afternoon, a 7 P.M. appointment to look at a loft for sale and then back to the office to edit prototype pages for a new section our paper is creating called Personal Journal. One notch below these obligations churn the smaller concerns: I need to call my mother . . . my toenails are a wreck from two weeks in the salt water . . . neck hurts from six hours in a tiny coach airline seat. . . . Guiltily I watch another set of joggers chug by . . . could cancel lunch and sneak out to the gym instead . . . will think it over in the shower. . . .

The conditioner is nearly rinsed out of my hair, and I'm lingering too long as usual under the warm water, when the first plane strikes. And with that initial deep thud, the day's dependable order explodes into mental Polaroid snapshots and sound bites. Hearing the noise, like someone dropping a cauldron upstairs. Yelling through the shower curtains to see if Kathryn is okay. Finishing my shower. Figuring it was nothing. Getting dressed. And then a phone call from our friend Erle, who lives in a building nearby—"Didn't you hear it?"—telling us to look out our window. Looking out and seeing the smoke streaking from the first tower across the street, hearing people scream beneath our window. Turning on CNN and seeing the second plane fly across the screen, an instant before we hear the roar in real time over our heads. And then the slam of the impact across the street. A bit of panic now—knowing that two planes can't be an accident—throwing on our clothes, grabbing our wallets, reporters' notepads and cell phones. The little decisions we'll regret: I choose open-toed black sandals. Contacts already in,

Kathryn leaves her glasses behind. Running down ten flights of stairs, not even bothering to dead-bolt the door. Because of course we'll be home for dinner. Of course Kathryn's fourteen-year-old cat, Stoli, will be better off here than outside in the chaos. Minds still glibly tuned to the way life is, with its reason and pre- dictability, it never occurs to us to look back and register home, warm and alive with our presence, one more time.

FROM ERLE'S APARTMENT a block farther south, we watch the Pentagon in flames on TV. Listen to the newscasters try to make sense of all this. Then a colleague from work, one of our top news editors, walks by the window. We yell at him to wait, gather our notepads and then go outside, where we congregate together by the river. What's going on? we ask each other rhetorically, our voices oddly pitched. We are wired, full of questions, scribbling details in our little books. This is a news event. We do not know to be scared.

The first tower's collapse is invisible. What we see are the gi- ant smoke plumes pouring around the corner, cartoonish in their dervish, billowing balls. What we hear is that rumble, a deep, horrible, guttural noise as if the earth were growling. Thinking that maybe a bomb had gone off in one of the planes, I am sud- denly alone several hundred feet down the river. Did I run? I must have run. Blinded from the smoke, I make my way back north against the tide of moving bodies, calling for Kathryn and Erle until I find them holding the river's guardrail and calling for me. Then together we join the throngs fleeing south along the river, racing from something we can't see amid the abandoned baby carriages, high-heeled shoes and unshaven men scurrying by in their curiously patterned boxers, briefcases tucked under one arm.

When the second tower falls—again, the deepest growl— we're at the very end of the island, now nearly covered in what seems a relentless white dust storm and still ignorant of what's unfolding, although there is now a rumor floating about the lips of the fleeing: a piece of one tower may have broken off. Aircraft storm low overhead, and not knowing they are our own military, the noise is terrible. I start to step over the railing, ready to face the Hudson River rather than whatever that noise brings. Kathryn grabs my arm and pulls me back. The sky is so dark I can barely see her or Erle, who is still carrying his *101 Dalmatians* coffee cup.

Nearby, a food vendor's cart is overtaken by the panicked mob seeking water, Pepsi, peach Snapple, liquid of any kind to wash off the white ash raining down on their faces, glasses and contact lenses. When the cart owner tries to collect money for the items, a burly stock exchange trader—his NYSE smock drenched in sweat—throws him to the ground.

"You do not charge anyone for anything," he screams. And so the looting is sanctioned.

I start talking to God. In the *oh-yeah-remember-me* way that only happens when I'm scared. I look up toward where the blue sky was an hour earlier—and I assume God is in fact up there somewhere in all this—and I manage to meet the occasion in a way that is spectacularly unpoetic. I promise that if this turns out okay, things will be different . . . I'll be different. Be better . . . change somehow. I'm wasting God's time because I don't even know what I'm trying to say. Still, I keep making promises anyway. And then those vague assurances sail up into the dark heavens to wait their turn amid the thousands being made by everyone around me.

A moment later, I slip into the crowd and steal a bottle of lemonade.

PACKED TOGETHER, THE three of us inch along the seawall, the smoke still pouring down from the Trade Center site a few

blocks north. The masses seem to be funneling into a nearby un-
derpass that leads to the FDR Highway. We hesitate to follow
them, fearing getting stuck under the earth. Then across the
street, in the opposite direction, a man begins pulling away in a
white truck.

Quickly, we scramble over road barricades, dodging the bare-
foot mothers clutching their children, filthy businessmen tapping
fruitlessly on their cell phones. Inside the truck is another food
vendor. It is Erle who knocks on the locked passenger door.

"Can we please have a ride with you?" he yells into the closed
window. The driver hesitates. All around him, people are fleeing. He does not want his truck overrun.

"Please, sir," Erle pleads. "There are only three of us."

Waving at us to be quick, the driver unlocks his door, and we
climb inside and crouch amid his still hot stove burners, bagels,
eggs and boxes of soft drinks and bread.

The driver rolls over a median and there is a moment of uncertainty—there's no clear path—and then he somehow navi-
gates his way onto the highway. His name is George Apergis, and
as he inches along in the smoke muttering to himself, the three
of us pass out Gatorade, water and soda to the streaming lines of
citizens making their way uptown on foot. We are glad for the
task, the sense of purpose it offers. Some people, clinging to custom, offer him money. George, who has a string of garlic hanging from his rearview mirror for good luck, won't take a cent
from anyone. We automatically scribble words in our notepads
as George talks: "Saw someone jump . . . inside they're dead,
outside they're dead . . . twenty-five years I've been down
here . . . try to help the people. . . ."

An hour later, George deposits us on the Upper East Side,
where men in still-crisp tailored suits stand clutching lattes and
staring at the filthy ghosts streaming off the highway into their
sensible neighborhood. Children and nannies point at the smoke

in the distance. Radios blare from parked cars; beauty salons have locked their TVs to CNN. Women mill about in pointy shoes, sniffing—*what's that awful burning smell?*—then realize the scent is coming from us and turn away. We look downtown, past the Empire State Building, toward home and work, now invisible in the haze. In this collapsed capsule of time—before we find a TV and see the horror we escaped, learn how we are among the luckiest down there today—terrorism is still, as far as we know, someone else's problem. Terms such as *Ground Zero, bin Laden* and *sleeper cell* are unfamiliar jigsaw puzzle pieces about to get fastened into context in the upcoming days.

And so we start walking, strangely euphoric, adrenaline pushing us toward a celebratory sense that yes, something very, very bad has happened, but here we are reeking, filthy and alive. Erle even has his *Dalmatians* mug. Our friend Sally lives nearby; we will go there, if we can remember which street. Seventy-third? Yes, that's it. We'll contact the office. File our notes. Try to reach our parents. Wash our faces. Figure out how to get Stoli the cat out of the apartment.

Today and tomorrow are all we think about. It does not occur to us to wonder where we will go after that.

WHEN I WAS very small, the safest place I knew was the bottom bunk bed of a trailer at my grandfather's old fish camp. The narrow structure crouched amid corn and soybean fields off a tiny slice of North Carolina's coast. It cost six thousand dollars and was our family vacation home until I was seven. My Granddaddy Bill, a Spanish professor, called it Pettiford, which roughly translates into "little river crossing." We didn't have much money then, but I didn't know it because Pettiford had everything. A view of the water. And a screened-in porch where you could stack sandy seashells and starfish without worrying if they started to stink. The neighbors' dog, Bandit, came right to the trailer door for day-old biscuits. Friends dropped by to visit without calling first. And amid the white oaks and bay laurel bushes hung a tire swing where snakes sometimes slept and had to be beaten out with sticks. I was never bored.

These are the last days I remember being content with what I already had.

2

the barman

Come back in the mornings.

There is a legend that describes how members of the Algonquin Indians were fated to wander the countryside until they found a river that flowed in both directions. Until such time, it was foretold, they could never settle in peace.

An impossible quest, it seemed, until they landed on the shores of a body of water whose heart was in constant play between the freshwater mountain lake, where she began, and the salty sea, where she ended. Each day, the ocean's tidal pulse made its play, rushing north with a mighty flood current that reached nearly half the river's 315-mile length, only to turn hours later in an ebb current heading south. For much of the year, the ocean dominated the quest, pushing its heavy salt front more than sixty miles upriver. But in the spring, helped by rains and melting snow, the mountains staked their claim and the freshwater runoff held the salt front back closer to the sea. Though fated never to choose, the river's blood was made rich from the dueling, and she supported some two hundred species of fish and a hundred and fifty species of birds.

The Indians named the river Muhheakunnuk, which means

"great waters in constant motion," or more loosely, "river that flows two ways." In centuries to follow, she would come to be called by a new name—the mighty Hudson. It was here on her shores, in a valley where the river ran deepest between high lands, that the tribes finally ended the search and made their home.

To ME, THE ivy-cloaked stucco house doesn't look like much from out front. A giant rusting metal and glass pay phone booth and a few scarred wooden benches occupy its front stoop. The listing green mailbox says GUINAN in peeling letters. A storm door is propped open, revealing a modest general store inside. Way in back, in what our friend Jessie promises is a bar, I see shadows of figures moving around.

"Trust me," Jessie says, drawing hard on her cigarette. "You guys have gotta see this place." She shifts her thin body impatiently in the cool autumn air.

Kathryn and I hesitate. It is midafternoon and we are due back in Manhattan soon to meet with a real estate agent. The rest of September and all of October have vanished, and now it's November 1. The *Wall Street Journal*'s offices next to Ground Zero are still shut from damage. Our own apartment building, which sustained the worst injuries in the residential complex, is also closed, but the government is letting residents gather their soot-covered possessions. No one is sure how long repairs will take, which means we need to find somewhere semipermanent to put them.

I look at my watch, anxious about the time, but not wanting to hurt our friend's feelings. I'm not even sure where we are— somewhere north of Manhattan's George Washington Bridge in a place called Garrison near where Jessie and her husband, Joe, the *Journal*'s art director, own a country house we've been staying at. Currently, we're standing on a river landing of some sort beside a train depot, and there are trees everywhere craning in toward the tracks as if to eavesdrop on our conversation. On my right, the Hudson River bends sharply, so I can't quite make out where it's going. The cliffs along the river's edge rise up haughtily, foliage spilling toward the water in fiery blankets. Wherever we are, it's a long way from steel and glass civilization as I know it.

We can't stay long, I tell Jessie.

"Just one beer," she promises, already moving toward the door.

THE MEN GROW quiet as we walk into the bar.

They are spaced apart on a handful of stools in a sprawl of plaid shirts, jeans and work boots, and there isn't really room for the three of us to sit together. So we stand awkwardly in the doorway, squinting to read the hand-scrawled beer prices on a yellowing piece of notebook paper tacked to the wall: Ballantine Ale: $2.75; Schaefer: $2.25; Murphy's Irish Stout: $3.75. . . . We are the only women in here.

The man seated behind the bar looks to be somewhere in his seventies. He is small and cheery, with red-rimmed, blueberry-colored eyes. He watches us, amused, it seems. Then he turns to a ruddy-faced fellow standing near the end of the bar, fixed there like a tree trunk, and jabs him with a few fingers.

"Jesus Christ, Mikey," he announces, his accent thick and clearly Irish. "Why don't ya move over so these ladies can get in here?"

Then he turns to us. "And what'll ya have, luvs?" he asks. "My name's Jim. Jim Guinan."

WE SETTLE ONTO stools at the end of the bar while the bar-man pulls our beers from the cooler. Jessie lights another ciga-rette. Kathryn checks messages on her cell phone. Meantime, I take a few quick sips of my beer and steal a look around.

To the left is a stone fireplace with a little pile of white ash, probably left over from last year. A green neon Budweiser sign in the shape of a shamrock dangles from one window. The scratches in the red bar top are deep, revealing scarred wooden planks underneath. Plastered on the mirror above the bar are law enforcement badges of every sort: United States Marshal, FDNY Engine 73, San Francisco Police. There's also a white road sign with an arrow and the words OFFALY 1. The smell inside the pub is something sweet and familiar, of sugared doughnuts, bread, beer and muddy marsh. In another room, I see a kitchen, clean dishes stacked next to the sink.

Whose place is this, anyway? I ask finally, breaking our silence.

The barman smiles. "Why, it's mine, of course, dear. I rent it, run it and live upstairs."

"And you've had it a while, right?" Jessie says offhandedly, trying to make conversation. The other men chuckle.

"Oh," the Irishman says. "About forty-two years now."

THE BARMAN'S STORY unfolds something like this. He arrived in New York City during the spring of 1957. The third child in a line of thirteen Irish Catholic children, James Guinan—no middle name—was searching for the American Dream and thought Manhattan seemed as good a place as any to start. He paid $708 to bring his wife, Peg, and their four young children, John, Margaret, Jimmy and Christine, over on the *Queen Mary;* that fare bought them a single room on the D deck with the ship's crew. The youngest, Christine, was only fifteen months old and John, the oldest, was five. The kids shared bunk beds and played cards throughout the six-day journey.

"It was a grand time," Jim tells us.

With six of them to feed there wasn't a lot of extra going around, and when Jim got off the boat at Pier 90 in New York City he weighed a scant 143 pounds. Work had been hard to come by in Ireland, he tells us, which was why he and Peg had been living in England and given birth to their children there.

So your kids are British? one of us asks brightly. I fear it was I.

"For the LOVE OF GOD," the Irishman bellows, causing the three of us to shrink back on our stools. "They might be born there but that makes no difference, because they're Irish." Standing at his tallest, I estimate Jim reaches only about five feet six inches. But the fierce look in his eyes ends the conversation then and there. I make a mental note never to confuse an Irishman with a Brit again.

They arrived at midnight and slept on the boat. The next morning, a cousin sent a car and they loaded up their luggage—the trunks were bigger than the four wide-eyed children. The family drove to a fifth-floor Bronx walkup apartment on Webster Avenue that belonged to Peg's sister. After helping his kids navigate the stairs, Jim stood alone at a window and stared down onto the noisy street below. Sirens roared in the background. He couldn't smell the wind. There wasn't a bit of green anywhere in sight.

"It was my first look at the city," he tells us, cocking his head sideways. "And 'twas my last."

The phone rings somewhere out in the store. "Excuse me, luvs," he says. "I'll be right back."

EVERYTHING IS STRANGELY quiet, except for the static of the VHF radio nearby and Jim's voice in the next room. Outside the window, the river shimmers in the sun, her surface unruffled. Leaves rain silently upon the yard in paper-thin hands of orange, rust and gold.

Leaning against the wall, I sip my beer, my mind drifting through the blur of events that have inadvertently led us to this stranger's pub.

On September 12, the day after the attacks, Kathryn and I walked back to our apartment building, evading police lines and National Guardsmen, in an effort to rescue her cat, Stoli. In the chaos, it had taken us nearly two hours to cover the few blocks to our home. The security forces never stopped us because we looked like medics, clad in oversized white shirts borrowed from our friend Sally's husband and wearing face masks around our necks. Upon reaching the end of the island, we cajoled two young National Guardsmen into escorting us past police to the apartment. Covering the riverside promenade like snow was a thick layer of whitish ash—much of it pulverized, save an annual report page here, a charred family photograph there. When we finally reached the apartment building, we could see that our tree-lined street in Battery Park City was strewn with monstrous twists of metal from the twin towers, pieces of wreckage being moved to look for survivors; the *Wall Street Journal*'s offices across the street, battered but still standing, were dwarfed by machinery and smoke. At this point, the boyish-faced guardsmen became polite but firm: they could accompany us no further.

"Building is unstable," they said. "We're under orders to stay outside."

So from there, we were on our own. Inside, we climbed with candles through the debris, up the empty darkened staircase and into the tenth-floor hallway, where doors had been busted open by firefighters looking for stragglers the day before. Our door too was ajar, and our hearts sank, thinking the cat must be gone. But when we pushed it open, there sat Stoli in the middle of the living room floor, a wide-eyed black-and-white ball of dusty fur. She looked at us as if to say, "Where have you been?" And then she meowed.

We grabbed her, our passports, a change of clothes and comfortable shoes and got out. For two weeks, the three of us camped out at a friend's office that was housed in a small apartment; we slept on the floor, Stoli now preferred the security of a closet. When it became clear that our own apartment would be closed for the foreseeable future, we scoured the *New York Times* real estate pages. One day I answered an ad for a West Village apartment rental that was owned by an eccentric, intense writer named Bob Rosenblum. Bob was affable but inquisitive—*Why are you moving? . . . Oh, I see. Well, were you home that day?*—and after about ten minutes of chitchat, he insisted I forget about renting his place and come meet his neighbors, a certain Linda and Lawton Johnson, who were going on vacation for a month in Italy. We all talked for a while in the Johnsons' living room; I petted their cat, Sam, and told them about Stoli. A month earlier, the situation would have struck me as odd, even suspicious. But this was an unprecedented time for New York, the city's cynical nature temporarily suspended and replaced by a notion that for once we were all in something together. And so it didn't seem so shocking when later that night my cell phone rang and it was the woman, Linda, whose home I'd left only hours ago.

"We'd like you to come live in our apartment while we're gone," she said matter-of-factly. "Keep Sam for us. Come as soon as you can."

And so it was that October ticked by in these kind strangers' home. In the meantime, we looked for a new place to rent. Having been a couple for two years, wherever we were going, we were going together. We started in New York City, but eventually expanded our search to Pound Ridge, Princeton, Tarrytown, Hastings, Nyack, Montclair . . . until the foreign-sounding towns blurred together as did the procession of upbeat real estate agents—Pat, Whitney, Herb, John. And suddenly the Johnsons were due back and we still hadn't found a place to go.

That's when we'd taken a break to visit Jessie and Joe up here. And now here we were sitting in this ramshackle bar in the middle of nowhere.

I'd known places like this, long ago when I was a kid. They had dotted North Carolina's rural coast, their collective space a surrogate country club for my young parents, who never quite fit into structured social life in our suburban hometown a few hours west. Smelling the river's scent slipping in the windows, I felt an unexpected tug at memories buried deep, hazy from neglect.

"Another round?" Jim the bartender asks, interrupting my

thoughts. He walks back into the room, limping slightly, I now notice. "Another round" isn't exactly a question, as his hand is already reaching into the cooler, his stance bending back toward his story. "And this one's on me."

Jessie lights another cigarette. Kathryn stares out at the river. Warm from the close quarters, woozy from the smoke and beer, I slouch deeper on my stool. And as the barman continues his tale, there in the midst of so much that is foreign, I find myself strangely calm.

AFTER A FEW hours in the Bronx, the family kept moving, Jim explains, winding their way north into the lush Hudson Highlands—this strikingly rugged fifteen-mile stretch of land where the Hudson River slices below sea level through the Appalachian mountain chain. It was a fair April day when they pulled into the tiny, secluded town of Garrison, near where another of Peg's siblings lived.

The heart of Garrison, as it turned out, was this river landing. Back then, mail was still delivered by train and carried across the tracks by cart. Forson's general store sold lamp chimneys, shoelaces, spades and pitchforks as well as the newest brands of paint, soup or cheese. Coffee was ground to order by hand in a large red mill with an eagle on top, and National Biscuit Company cookies were sold loose by the pound from large bins. Up the road was the Garrison Reading Room, a space that had moonlighted, through the years, as an athletic club, polling place and garage for the single engine of the volunteer fire department.

"Oh, the folks around here have always known how to make do," Jim says. For instance, he asks, did we know that the multi-level brick house up the road was once home to a hotel and bar whose owner kept business going during Prohibition by disposing of whiskey bottles through holes cut in the river ice. "Ah, that's a smart lad he was now."

Prohibition? I interrupt, my tongue loosening with the drink. So this town must be pretty old?

Mike and his companions smile, glancing at Jim.

Uh-oh, I think.

"For the LOVE OF CHRISTMAS," Jim bellows, leaning on his elbow and turning his head so one eye pins me. "Luv, would ya turn around and look outside that window. Right there." He jabs several times behind me with his pointer finger. "Do you see those fortresses on the other side? That's the United States Military Academy at West Point. And who do you think built that?"

I keep my mouth shut.

"George Washington, that's who. You heard me. He was here with his troops. Strung the great chain across the river right out there during the Revolution to keep the goddamned British from making it any further up the Hudson. For Christ's sake, they've all come through here at some point. President Lincoln, Robert E. Lee, even that traitor Benedict Arnold."

I can't remember the exact details of what Arnold did, but now doesn't seem like the time to ask.

"How old is Garrison, you ask, dear?" He leans forward dramatically, placing a hand on my shoulder, his voice gentler now. "I'll ask you—how old is freedom?"

"Now," he says, settling back behind his bar, "as I was saying, *I* didn't get here until 1957. . . ."

IN THOSE DAYS, Jim says, a man could make a decent living on the river landing. But he might also find some peace. Working as a carpenter and helping to build a new dock earned enough pay for him to rent one of the landing's colorful old row houses built during the 1800s. Their first home was half a duplex just up the road, a tight squeeze for the family of six, but there still was plenty of richness as far as Jim was concerned. A fine peach tree and maple stood outside the front door. When rain came, you could smell its approach dancing off the river before the first drops fell. At night, there was quiet, the kind that made for deep sleep. Children fished, biked and swam without supervision but still looked their elders in the eye and said yes sir and no sir.

Garrison, Jim Guinan decided, was about as fine a place as any to call home.

———

KATHRYN GLANCES AT the clock on the wall and then looks over at me. It's time to leave. We stare at each other a moment, then she shrugs. And I'm grateful.

Because more than anything in the world, I do not want to get up off that bar stool.

"So how'd you come to own this place?" Kathryn asks, turning off her cell phone in case the annoyed real estate agent calls. She drops it into her bag.

Jim gives her a stern look and reaches back into the cooler. "Be patient a bit, will ya, dear?" he says as he sets out another round. "I'm getting to that."

———

BY THE TIME he finishes his story we have made friends with this guy called Old Mike and completely missed our appointment with the Manhattan agent. As the afternoon makes way for early evening, Jim speaks of castles and robber barons and other crazy things that apparently have something to do with this little town. His wife, Peg, is gone now, the barman concludes, passed away Easter Sunday in 1988. Coffee is brewed, not instant. And the bar has moved a dozen feet from the store into this room.

"Otherwise," Jim says proudly, with a small cough, "not a lot has changed since we started things."

Finally, with darkness settling in and the lights of West Point blinking on across the river, we pay for our beers and prepare to leave. At the doorway of the bar, I shake the barman's hand and reluctantly bid him good-bye.

He gives a nod and a wink.

"Come back in the mornings," he says. "I'll have your coffee ready just how you like it."

Ten days later we move to Garrison.

AT PETTIFORD, MY grandfather liked things a certain way. He kept a cooler full of Pabst Blue Ribbon, a fat green tackle box with purple rubber worms, and multiple glass jars full of things whose purpose was clear only to him——broken lure weights, a brittle skate's fish-egg purse and a crushed Cadbury Cream chocolate egg. In his glove compartment was a soft black change purse filled with dimes and nickels——his emergency money. My mother's father was a man of habits: Kentucky bourbon neat, White Owl cigars, M*A*S*H and a stubborn inability to show affection for those he loved most. Animals and small children were the exceptions. He named his fishing skiff after me, the only grandchild, dubbing it the SS Wendy B. In the evenings, he would sit in a rocking chair for hours after dinner staring into the dark river. He was not to be interrupted but allowed me to sit quietly on his lap. He could always sense a storm's approach.

When the rain came, it would beat on the tin roof like a metronome. Cocooned into the bottom bunk, I'd fall asleep under a heavy wool blanket listening to my parents laughing low in the other room, and know that come morning, with the water calm and flat and the sun shining again, my grandfather would call me to watch the blue heron stalking minnows off our pier.

3

the barman's children

. . . it stands for Ass, Tits and Temper.

Someone once described Garrison as a "state of mind." And her physical boundaries are suspect, slinking around small mountains and through treetops, down to the river, back across broad fields and encompassing fewer than twenty square miles according to one official count, but seeming far bigger or smaller depending on the season's tree cover. Population: roughly four thousand.

In late 2001, Garrison still has no mayor, no town water and actually isn't big enough to qualify for official town status. Instead it is a hamlet, a mere snippet of life inside a broader swath of rugged land dubbed Philipstown—a throwback to the late 1600s when the king of England was doling out land patents and gave a prime chunk to one wealthy New York merchant named Adolph Philipse.

Garrison's official name, as much as anyone can agree, comes from a Garretson family out of Holland whose descendant Harry ran a river ferry here in the early 1800s, although this late date is not where the hamlet's story begins. Every so often a blue and gold sign pops up along the roadside to tout the escapades of Washington's Revolutionary War forces and remind anyone, lest they forget, of the place's front-row seat at our nation's birth. True to Jim Guinan's word, one of them reads: "At Beverley Dock, at the foot of this lane, Arnold, exposed as a traitor, fled by boat to the British ship *Vulture,* off Croton Point."

There are also, as the Irishman promised, "castles." Modeled after Europe's greatest architecture, they stand high in the trees, one a towering fortresslike granite mansion with balconies and roof perches that wrap into the landscape, and the other an off-white concrete empire echoing Spain's Alhambra. Both date to the late 1800s when rich railroad barons and financiers roamed the lush Hudson Highlands, earning the region the nickname "Millionaires' Row." Today, the scattered residential homes are a mix of modest and lavish tucked together amid rambling stone walls, hay fields and dirt roads. Garrison is a fifty-mile train ride from New York City's Grand Central Terminal and one of the last stops on the Metro-North line bound for Poughkeepsie.

Driving down the main road, you get the sense that somewhere along the line, somebody locked the gates to the commercial world and said, Thanks, but no thanks. Fast food and mass merchants are nonexistent, the closest Starbucks or Wal-Mart a twenty-minute drive. To live here is to escape into the calm eye of the strip-mall storm that clogs the cities immediately north and south. Deer and foxes roam freely, power outages are common and directions are dispensed in a kind of local code borrowed from the rugged terrain: *through the Rock Cut . . . over the Goat Path . . . after the Dip. . . .* Old-timers sport limited edition license plates stamped with Putnam County's 9X or PC series. Even the celebrities seeking refuge here through the years— Sylvester Stallone, singer Don McLean, actors Kevin Kline and his wife, Phoebe Cates—have always just been quietly absorbed into the landscape with everyone else.

The hamlet slipping into winter is a gentle but palpable occurrence. As final fair warning, the leaves begin to glow until it looks as if someone has lit a warm lamp across the entire valley. The sweep of Sugar Loaf Mountain, where American troops stood lookout during the war, turns more dramatic with its deepening palette of maple, hickory, oak and birch. Downey Oil's yellow trucks work overtime, their drivers topping off tanks, inspecting copper piping and pacifying the older folks who fuss and phone several times a day, worried their burners might fail in the first big storm. Wood is stockpiled, chimneys swept, snowplow contracts signed.

Meantime, the summer renters close up shop, anxious for city life where capable doormen and superintendents keep harsh elements at bay. Those who stay can watch as the landscape strips down, unloading her buxom summer weight until she is naked, angular, severely beautiful.

This is home on the eve of our arrival.

THE THREE-BEDROOM HOUSE we rent sits on several acres of land and shares a dirt driveway with one other home. Directions to our baffled city friends now include "turn right after the deer-crossing sign." Our trash collector, Louis Lombardo, carries giant dog biscuits in his white truck for the town's pups. At the closest movie rental store—Mike's Video Madness—the proprietors also do wash at the Lost Sock Laundromat in back, eight bucks a load. The first morning here, I open our front door, arms crossed against the chilled mountain air, and watch a wild turkey strut across the driveway with her babies and disappear into the thick brush.

Our neighbors are a Dutch woman named Jos (pronounced like "yo" with an *s*) and her husband, Walter, a sixty-two-year-old retired salesman for a high-end men's sweater company. Walter is a man of particular order and custom. He wakes at 6 A.M. to smoke a cigarette in his bathrobe on the front porch, skinny white legs visible through the woods between us, before going back to bed. He reads every page of his newspapers and scrutinizes the weeds in his yard with the zeal of a man possessed. Occasionally I catch him setting fire to white-headed dandelions with his Bic lighter, methodically torching each one while muttering, "I hate 'em. Hate 'em."

Walter's frugal streak is well honed, and he sees in us ready pupils, spoiled by the excesses of city life and in sore need of his wisdom. For instance, gas, he instructs, should be purchased only on Tuesday or Wednesday when it's discounted by three cents at the Gulf station on Route 9. Our neighbor's entire house is painted in Benjamin Moore Linen White, and Walter cannot bear the thought of using a finish other than flat. ("You can see EVERY flaw.") He has used the same paintbrush for seven years. "If you take care of it, it will take care of you," he says, wagging a long finger.

Walter's wife, Jos, is a horticulturist who plants oregano in our yard, digs up wild onions at a feverish pace and talks about how things have changed in Garrison, and not necessarily for the better, mind you. For example, when they moved here in 1989 there were still so few houses that she could garden in her underwear, which, she explains in her thick Dutch accent, "was really nice because Wally is too cheap to buy air-conditioning." A moment passes with her plunging her trowel into the earth before she adds: "But don't worry. I don't do that anymore."

I look at her, puzzled.

"Garden in my underwear," she says.

JIM GUINAN'S PLACE is just three minutes down the road, across the railroad tracks and on the river landing beneath a hill where New York's current governor, George Pataki, makes his home when he's not wrangling with legislators in Albany.

If a downtown exists, this Garrison's Landing is it. In September of 1609, Henry Hudson had sailed up the Indians' "river that flows two ways" in his high-pooped yacht the *Half Moon*. As his crew headed north toward the Highlands—her rocky cliffs high above their ship—the explorer bestowed the body of water with yet another name: "River of Mountains." Later the Dutch and other settlers came to live along the shores. As they cleared tracts along the River of Mountains, boat landings began to spring up to transport goods and people, and a thriving community arose in a place that had once seemed terrible and wild.

Subsequently, commercial enterprises of one sort or another have long occupied Garrison's Landing, although the nature of the clientele has changed quite a bit by the time we arrive. For instance, a performance theater occupies the old railroad station, finished in 1893 and once used by the railroad tycoons along Millionaire's Row. The multilevel building that once served as a hotel now houses an iconoclastic lawyer, a wine merchant and a yoga studio. Looking closely one can see the name Vandergelder etched into its glass doors—a relic from when 20th Century Fox descended here to film *Hello, Dolly!* starring Barbra Streisand as the devilish matchmaker Dolly Levi in pursuit of skinflint Horace Vandergelder, played by Walter Matthau. (The two stars didn't get along too well, apparently; residents still recall when Streisand handed Matthau a bar of soap and demanded he wash his mouth out to purge his foul language.) Other landing tenants include a quirky Australian rare book dealer and an excellent masseuse named Georgia, who advises clients to put powdered ginger in their socks to improve circulation during the winter.

The Guinan country store and pub is the last structure at the southernmost tip of the depot's square. For a week, I pick up a cup of coffee there before driving to work and hope to see Jim again, but he's not behind the counter. Instead, a younger man and woman seem to alternate the morning and afternoon shifts. They look to be in their late forties, with pale skin, attractive faces and tired eyes. The guy has a white handlebar mustache, a good-natured face and a well-built upper body that suggests he doesn't work behind a desk. The woman looks a little like Meryl Streep and is fine-boned, serious, with a long, elegant neck and a mane of thick blond hair she pulls into a ponytail. They seem to know everyone who comes through the store.

On my fourth morning there, I introduce myself to the woman.

I'm Wendy, I say. The words come out brightly. I just moved to Garrison, I tell her, and extend my hand across the counter.

She looks down pointedly at the shiny metal slicing machine where she is clutching some kind of greasy meat—looks like roast beef. Her fingers are covered in dark juice and slime from the meat. I awkwardly retract my hand.

"I'm Margaret," she replies. Nothing more.

I try again. The man who lives here, Jim, I say, hoping to impress her by dropping his name. Older guy. I don't see him out here anymore. Do you know what happened?

She turns off the slicing machine and puts down the meat, wiping her hands with obvious distaste on a white towel. "That's my dad. He's been sick—some complications with diabetes. My brother and I help him so he can get off his feet for a few hours."

Oh, I'm sorry, I tell her. Again, she offers nothing else. Then I notice her sweatshirt, which says AT&T in bold black letters. Ah, I think, some common ground.

So you work for AT&T? I ask, leaning against the counter in what I hope is a casual pose. Lot of turmoil there right now, huh? All those layoffs? And then, to let her know I really understand what I'm talking about, I add: I write for the *Wall Street Journal*.

Finally she meets my eyes. She doesn't smile.

"I'm in law enforcement," she says, lips tight. "I wear this sweatshirt so when the guys at work give me a hard time, I can tell them it stands for Ass, Tits and Temper."

MAGS OR MAGGIE, as they sometimes called her on the force, had spent sixteen years at the police department of a neighboring town, most of it as the precinct's only woman. When she first became an officer, she knocked on the door of someone who'd called for assistance, and he looked at her skeptically and sneered: "Yorktown doesn't have female police officers." Margaret opened her jacket, flashed her metal piece and glared back: "Yeah, well, this badge and gun says I'm one."

The eldest Guinan daughter learned to shoot as a teenager in an apple orchard owned by the local building inspector. He set up railroad ties with targets; she nailed them with a .32-caliber Smith & Wesson. These days she's a detective and carries a heftier standard issue 9 mm Smith & Wesson on the job and a smaller and lighter Glock 9 mm off-duty. (The same building inspector sold it to her for four hundred dollars, a pretty good deal, she notes smugly.) When she applied to the force, she was sent to a DWI seminar for officer sobriety test training. The instructor told everyone to practice the tests in the field and gave them a week. Margaret marched into her dad's bar, ordered the patrons to put down their beers and made everyone walk a straight line and stand on one leg; she ended up driving half of them home. She also turned her assignment in the next day. Now the detective mostly handles sexual abuse investigations, particularly those involving children, but she's got to be ready for anything: robberies, burglaries, high-speed chases.

Kids in her jurisdiction know her by sight and holler "Hello, Detective Maggie" when they spot her walking around without her jacket, gun strapped to her hip. Margaret, however, has another popular nickname aside from Mags or Maggie, and that is "bitch," which is what a suspect inevitably calls her when she's arresting him. This doesn't bother Margaret in the least. "How did you know my name?" she'll deadpan while tightening the cuffs and pushing the culprit's head down into her patrol car.

The detective has a shamrock tattooed in the colors of Ireland's flag on her right calf, and when I ask if it hurt, she looks me in the eye and says she has a high tolerance for pain.

OKAY, MARGARET TERRIFIES ME.

Still, this place is the only game in town, so I keep coming back, hoping to see Jim eventually. And the more I hang around, the more the detective seems to at least tolerate me.

Despite her toughness, the officer is lovely in a complicated way that comes and goes with the clientele. There's the quiet, scruffy man with a squinting eye who always comes in for a Bud Light around 11:30 A.M. Despite a roughshod appearance, his shirt is always tucked in and his laugh is something warm and infectious. Margaret speaks gently to him, charging $2.25 instead of the usual $2.75 for his Bud Light. "He used to drink Schaefer's, which only costs $2.25," she explains after he leaves. "But then he got sick and the doctors told him to switch to light beer. Dad thinks it's only fair to charge him what he paid before."

And the moment a child steps into the store, I'll see the detective's eyes soften and suddenly she's handing out free pretzels and kneeling beside them at the candy counter to help with the difficult decisions: Tootsie Roll or SweeTarts?

Margaret is not married and at age forty-eight has no kids of her own, which isn't likely to change, she points out, while pulling down fourteen to sixteen hours a day between helping out at the store and working on the force. Plus, she adds, where's she going to meet someone? The usual spots—like bars—aren't much fun because there's usually someone she's arrested there. Margaret never wanted her parents' life—up at 4 A.M., on their feet until late in the evening. No thanks, she says while emptying the ashtrays and wiping down the bar. Spending her childhood

with the town streaming in and out of their house every day—
that had been plenty. She wanted a steady paycheck signed by
someone else, which is in part why the regimen of law enforcement suits her so well.

But Margaret still arrives here every day around 8:30 A.M., and
as far as I can tell, is the one running the show. She tallies sales, orders new inventory, stacks soft drinks, restocks beer, slices fresh
tomatoes and lettuce, sweeps the floor, carries out the trash,
makes sandwiches and rings up bubble gum and cigarettes. She
changes her father's foot bandages and makes sure his lunch and
dinner are fixed.

Around 3 P.M., the detective goes upstairs into the room where
she grew up and puts on office clothes and makeup. Then she
kisses her father good-bye and her real workday begins.

Once, a customer asks Margaret how long her family has owned
Guinan's.

"Feels like about twelve thousand years," she says, sweeping
past him.

JOHN GUINAN IS different from his sister, instantly chatty
and extroverted. He sits in the bar some mornings between
trains, telling loose stories about whatever comes into his head.
John is full of aphorisms, which he uses liberally. Ask a question
without a clear answer and he'll tilt his head and ask you: "Why
is there a fence around the cemetery? Because people are dying
to get in." In John's world, a guy isn't merely cheap when he
constantly bums cigarettes at the bar; he either "squeaks when
he walks" or is "tighter than a frog's ass—and a frog's ass is wa-
tertight."

Ask John where he gets his sayings and he'll just shrug. "I
don't know. Dad, I guess." A pause. "Hey, you know why God
invented alcohol? So the Irish wouldn't take over the world."

I'm thirty years old. But he calls me kid.

John works full-time with the town's best arborist, a guy named Lew, taking down trees, pruning, ripping out stumps. He handles the early shift at the store, rising at 4 A.M. to get ready for the first wave of commuters catching the 5:09 train. He does things a certain way every day. Writes some customers' names on their newspaper in the top right corner with a black lead pencil and lines them up under a Reserved sign. He stacks coins and dollar bills on the counter so people can make their own change for papers and coffee and catch the train on time. Ask him about theft and he'll look at you and say, "This is Guinan's," as if that's all the answer needed.

John only uses one button on the rickety 1927 wooden Federal cash register—the one that opens and closes the drawer—and it requires a solid punch with the pointer finger. When the drawer gets full, John transfers the cash into rubber-banded packets of fives, tens and twenties and stuffs them temporarily into a wooden Corina Ole cigar box hidden in his dad's living quarters. He never uses a calculator. I never see him get a bill wrong.

"How do you do that?" I inquire once after he's quickly tallied a long list of items: $5.95 cigarettes, a 50-cent roll, two 90-cent bagels, a 75-cent *New York Times,* a 50-cent *Daily News* plus an orange juice and roll of Vicks cough drops thrown in at the last minute. "How did you learn to add in your head so fast?"

"Dad," he says again. "He learned to do it playing darts. And he made all of us learn."

I CAN'T FATHOM how John manages to get up so early and still work a full day outside doing manual labor.

Aren't you scared you'll fall out of a tree or something? I ask him one morning while he's changing into his heavy work boots and fleece pullover. I mean, you must be exhausted, I say.

He smiles and then pops a bunch of vitamins into his mouth. "Sure, kid. But you get used to it."

For nearly a decade before he began doing tree work, John had a job indoors at a General Motors car factory located thirty minutes south of Garrison. He was one of nearly thirty-five hundred workers who assembled parts inside the plant, which had been built at the turn of the twentieth century to produce steam-powered cars. The factory, acquired in the early 1900s by GM, had seen its heyday in the 1950s and '60s when workers were churning out Chevrolet Bel Airs and Impalas. By the time John arrived in 1985, the operations were mostly dedicated to assembling GM's line of long-nosed minivans—a roomy family vehicle that would soon cede its throne to a jazzier breed of road hog: the sport-utility vehicle.

His wife, Mary Jane, meantime, delivered mail for the post office in the next town north of Garrison, Cold Spring. She and John had three kids, but despite their two incomes, hiring childcare help was an unaffordable luxury. Instead, the couple structured their schedules so one of them would always be home. On a typical day, John would leave the factory at midnight and arrive home around 1 A.M. He'd drink a glass of milk and try to sleep for a few hours. Mary Jane had to be at work by 6:30 A.M., so by the time John woke up around 7 A.M., she was gone. He'd hustle his two oldest kids, Sean and Kelly, to the school bus, then load his lawn mower and landscaping equipment into the back of his truck. The rest of the morning, operating on only a few hours of sleep, John groomed yards around town to earn some extra cash while carrying his youngest son, Casey, strapped onto his shoulders in a backpack. Around lunchtime, he'd head back home, shower and leave by 2 P.M. to punch back in at the factory. During these years, he and his wife communicated mainly by notes, passing each other only in the night as they lay side by side in bed for a few hours.

It was never the labor or long hours that bothered John; the pay and benefits at General Motors were good, and his logical mind helped him do well on the job, which entailed welding, fitting trim, assembling transmissions and whatever else his supervisors asked. But for a man who'd spent most of his youth freely roaming the wooded shores of the Hudson, the confinement of being indoors so long gnawed at him. On his break—usually around 8:30 P.M.—John would hurry to a nearby deli, grab a can of Foster's beer and go sit by the river with his sandwich in the dark. He'd study the moving riverboat shadows, listen to the slap of water on the shoreline and imagine a time when he could be outside as much as he wanted. Then, at the hour when most families were washing dinner dishes or settling down to watch their favorite TV show, he'd head back inside the factory.

Which is why, he explains to me, he doesn't mind going to work all day with Lew after getting up so early to open the store. "I love my job outside," he says, rubbing Vaseline across his face to protect against the wind. "I mean, I *love* my job."

JOHN'S FATHER STARTED having problems with his feet about five years ago. But he never said anything to his children. Never said anything to anyone as a matter of fact. At the time, Jim's youngest daughter, Christine, and her husband—who both live in Florida now—were helping her dad run the store. Christine noticed her father limping and called Margaret, who immediately drove him to a local doctor. The doctor spent a few moments alone with Jim before asking Margaret to join him in the examining room.

"I want you to look at this," he said.

Jim was sitting on the table with his pants folded up to his knee and there was a deep wound in his left foot. What had happened was this: Jim had adult-onset diabetes, which he had long controlled through his diet. He also suffered from a connected condition called neuropathy that caused a loss of sensation in his feet. The lack of feeling meant he didn't distribute his weight evenly when he walked or stood for long hours behind the counter or bar. As a result, calluses would develop, but because he couldn't feel them, Jim would just keep trucking along and making things worse. Eventually, the doctor told Margaret, her dad had worn a callus on his foot that bothered him so much that he shaved it off with a razor blade, leaving this wound.

"It was a nothing that became something," John says.

Over the next few years, their father's health worsened. He began taking diabetes medications. Still there were more infections, more trips to the doctor. A sore would heal, and then Jim would get back on his feet to work in the store and things would flare up again. In early 2001, Jim picked another callus off his foot and ended up back in the hospital. This time, the infection set into his bones, so they hooked him up to an intravenous antibiotic feed. But the second toe on his left foot was red, swollen and oozing. When the doctor moved it slightly, it snapped. She scheduled him for surgery and said to hope for the best.

The barman lost the toe but kept his foot. That's when John and Margaret began helping out more. Margaret switched her schedule to permanent night duty and used up all her personal days at the store. John began his 4 A.M. wake-up routine, promising his wife this was only for the short term. Neither he nor Margaret took a cent for the time they spent at the store. "We figured we'd be there for a few weeks while he healed," John recalls.

This was nearly a year ago.

THERE IS A cousin in England, Kathleen, who is trying to get a visa so she can come and help out at the store. But there's a backlog because the INS is scrutinizing everyone more carefully since the terrorist attacks. Christine, Jim's other daughter, is now firmly settled in Florida with her husband and children. The fourth Guinan child, Jimmy, has lived in San Francisco since the late 1970s, and no one is ever quite sure when he'll show up for a visit. John talks to his dad about hiring someone to help out so he and his sister can get a break. Even puts a Help Wanted sign in the window—which his dad promptly removes, and John doesn't argue. This is, after all, a family place. Their father sleeps upstairs. Money is in the house. Plus, and no one needs to state the obvious: this is Guinan's. If there's not a Guinan here, well then . . . yes, well then . . .

So they talk. But nothing changes.

THE CUSTOMERS ASK John, "Hey, why don't you take over for the old man?"

He smiles tightly. "What, and have to put up with you all the time?" He means it as a joke. But if you look carefully, you can see that his jaw is clenched, and he's sorting the coffee filters a little faster.

Well, why not, I ask him one morning, during a lull between trains. He sometimes dreams out loud about adding taps in back for Guinness or upgrading to commercial-grade appliances so the store could offer hot breakfasts. Why wouldn't you think about running this place? I ask.

He looks away, rubs the bridge of his nose between the eyes. "Once," he says finally, "a long time ago, I talked to Dad about it." That's all he offers. Then he walks into the kitchen to chop up old bread for the seagulls hovering out back for their breakfast.

Another thing about John. When he doesn't say much, there's usually more to hear.

STOLI STOPS EATING and drinking. The skin beneath her white fur turns yellow. We take her to a local vet up here, who explains that the cat's liver isn't functioning right. It could be because of what she breathed the day of the attacks, the doctor tells us, or a combination of that and the stress of moving around so much; there's no way to tell—except through an autopsy.

With the doctor's guidance, we try to hydrate her at home, pushing a needle several times a day through the loose skin of her neck and letting the fluids drip into her while she sits on our lap. We feed her a paste loaded with nutrients to try and get her strength back. But nothing seems to work, and so we finally have to check her into the hospital. For a few weeks after work, Kathryn and I navigate along unfamiliar dark country roads to the tiny clinic where we sit petting Stoli amid the din of dogs barking in their kennels. By Thanksgiving, she hasn't improved, and the doctor asks permission to perform a last-ditch operation during which she inserts a feeding tube into our cat's stomach. A day later, Stoli is barely hanging on, and there's a decision to make. When we reach the hospital, Kathryn wraps Stoli in her jacket and the three of us crouch together alone in the examining room next to a cold steel table. A few minutes later a new doctor knocks on the door, syringe in his hand. He asks if we are ready. Kathryn nods.

A few moments later Stoli dies in her arms.

REPAIRS AT THE *Wall Street Journal*'s offices may take as long as a year, we're told. There is talk about asbestos and other substances in the buildings that must be cleaned out before we return. Our old work belongings—files, computer disks, photos, headphones, books, lamps, CDs, Palm Pilot cradles, a sweater left on the back of a chair—are cleaned and packed up by strangers. Then the boxes are numbered and sent to a windowless room in our corporate offices in South Brunswick, New Jersey, to be reclaimed.

The staff is eventually split up on a semipermanent basis: Reporters, including Kathryn, head off to a temporary office in Manhattan where they sit crowded like telemarketers but glad to be back together. Most editors operate from our corporate offices in South Brunswick. I'm in that camp, but can work from home a couple of days each week. I spend those early mornings at Guinan's.

Our managing editor, Paul Steiger, rallies the troops. Tells us he's proud of us and knows we'll all give 200 percent during this difficult time. And people do their best. Reporters cover their beats without crucial notes and research left in the office on September 10. Some of the South Brunswick crew, particularly the daily news editors, hole up in a sterile-smelling Marriott Residence Inn several nights a week rather than face the long commute. They phone their children and spouses to say good night, assuage strained family nerves and then ride on vans together, shoulder to shoulder, from the Marriott back into the office the next morning to do it all again.

Around the newspaper, around the city, around the country, everyone is trying to find their way back to some sense of normalcy, even though the rules of daily life, as we've all understood them, have just been chucked out the window for good. News reports show people packing face masks, flashlights and AM radios in their briefcases alongside hairbrushes, lipsticks and wallets. Some talk of driving a little faster over bridges, picking a less crowded time to go through a tunnel—maybe taking a different route through the train terminal. Just the act of making choices, a semblance of control, is what matters.

I do all those things. I also go as much as possible to the stucco house by the Hudson River. It is there, I sense, that I'll find my own way back.

"SIT STILL," MY dad told me.

We were in the Wendy B. *running nets just offshore of Pettiford where my grandfather waited back on the pier. It was dawn, and I was in my pajamas with a coarse blanket wrapped around my shoulders. The* bottom of the tin boat was slippery with flopping fish, which I kept poking with my little black rubber boots.

"Listen," my father said, cutting the motor. "Do you hear that?"

No, I said, not listening at all, and kept jostling the fish.

My father leaned across to the bow and wrapped his arms around me. "Shhhh. Stop. Now listen." I fidgeted for a moment, then relaxed in his grasp and looked toward the tree canopy where he was pointing. The only sound was the gentle cluck of current against our hull, and then—

whip-poor-will . . . whip-poor-will . . . whip-poor-will . . . whip-poor-will . . .

I looked up at my dad questioningly. He nodded.

"That's the bird I was telling you about. The one who knows his own name."

I tried whistling back to the bird—but it just sounded like I was blowing air. My father laughed and rubbed my head. "Keep trying, because someday you'll be too busy to hang out here just listening to the birds," he said.

That's silly, I told him, whistling again and again until I finally forced out a little sound . . .

whip-poor-will . . .

fitz and the

friday night parishioners

I've been no bargain over the years . . .

A Friday morning, winter's cold grip securing the land. The holidays have come and gone, and I'm back in the bar drinking coffee and hear the store door open and slam shut. A male voice greets Margaret just around the corner from where I'm sitting— "How are you?" Nothing unusual about the question, but the softness in her response makes me pay attention.

"Tired," she says. "Working here. Working there. Dad was back in the hospital for a few days."

"Yeeaaah," the voice says, "I heard that. I'm sorry. I'm not even going to give you a hard time today."

"Hey, thanks," she says. "Why don't you get yourself a cup of coffee? You aren't waiting around for me to do it, I hope."

"Never," he replies, and you can tell from his tone that he's grinning. "I know better than to mess with a DT," he says, using the slang for detective.

Another customer, a woman, interrupts their talk with a breathless question: "Excuse me but can either of you tell me, is it the *Post* or the *News* that's published by Mortimer Zuckerman?"

"Beats me," Margaret says, the old tiredness returning to her tone.

The words are out of my mouth before I can stop them. The *News,* I call out.

The man peers around from the coffee station into the bar, staring at me as if I were a curious artifact he'd discovered in his own backyard. He's tall, well built with a thin mustache and pale skin. His eyes are hidden behind dark aviator sunglasses and his dark blond hair, what's left of it, is cropped close, military style. The nose is sharp, thin, and his ears stick up straight, as if they were at attention. The entire demeanor suggests that of a man accustomed to commanding a room. He watches me for a moment until I feel myself begin to shift uncomfortably on the green stool.

"Who's this?" he calls finally to Margaret, his voice clipped with a northern accent. He hasn't stopped staring at me. Or at least I think he hasn't. He's still wearing the sunglasses.

"That's Wendy," she says, squeezing past him into the bar. "Wendy, meet Fitz. But don't trust anything he says."

"Wendy," he repeats, almost as if he doesn't believe her. He leans against the end of the bar, close now, with his cup of coffee. "Wendy what?"

For some reason I hesitate but can't think of a good reason not to tell him, so I answer reluctantly. Bounds.

"Wendy Bounds." He chews on the words, drumming his fingers on the bar. I notice his fingertips are swollen and strangely white. "Well, what are you writing there in your notebook?" he asks.

Instinctively I pull it closer to me. Just some notes, I tell him. I'm a writer. Before he can go on the offensive again, I ask, What do you do?

Fitz grins, says kind of out of the side of his mouth, "Well, if I told you that, I'd have to get rid of you."

Margaret sweeps past with a bag of new coffee filters. "Don't let him talk to you like that. He's a retired federal marshal."

"Yeah, well, don't go writing that down," he says gruffly but doesn't seem that upset she's told me. He takes off his sunglasses, revealing smallish blue eyes that quickly scan the room, almost imperceptibly, before settling back on me. "You're new to town." It's a statement, not a question. "Where do you live?"

I smile. Now if I told you that, I'd have to get rid of *you*.

He laughs; it comes out loud, like a nasally heh, *heh, heh*—but with an earnest smile that reveals more of the teeth, smallish and sharp, and he leans deeper into the bar, begins talking.

"Well, you're smart enough to have found the best place in town, I see. Like a time warp, isn't it?" I'm clearly listening to him but he elbows me as if to make sure I'm paying attention.

"Nineteen-fifties America frozen in time," he continues. "There are a few of these places I liked to go over the years. The Schlitz Inn in the Bronx; I think it's gone now. It had old stools with the Schlitz beer logo on them—"

He elbows me again before continuing.

"—and it was run by an old German lady. She made everyone's lunches one by one. There's also Izzy's downriver in Ossining—you know Ossining, right? Where the prison, Sing Sing, is? That's my hometown. And then there's Shannon's Tavern in Jersey City." *Elbow.* "Listen to me, I started going there when I was eighteen. You had the meatpackers in the morning and cops and FBI in the afternoon. One of the bartenders, he had a droopy half-shut eye. All the owners of these types of joints are in their seventies. Like Jimmy Guinan here at this place. Yeeaaah, but listen, they're dying off. Hey, you know, these guys are becoming obsolete."

Elbow.

AND THAT'S HOW I first met William Fitzgerald. As I got to know him, I came to understand that a guy like Fitz is the lifeblood of a place like Guinan's. He's the one who always has an opinion and will offer it up about damn near anything: politics, war, taxes, the *real* problem with women. . . . More important, if nobody else is offering up much interesting, a guy like Fitz will provoke them until they do—"Hey, what do you need a beer to relax for? If you relaxed any more you'd be a lounge singer at the Holiday Inn"—considers provoking his duty, in fact, just to keep things from getting boring. He's proud of it too. "First day I walked into Shannon's, I went as a guest, and within a half an hour, I'm having arguments with guys like I'd known them all my life, and they're calling me an asshole." He talks fondly of La La, another Shannon's regular with a knack for stirring the pot. "La La, he'd say stuff like 'The best president ever was Jimmy Carter' to a bunch of conservative military guys. He was just digging at everybody. Like when he'd give a toast to the Queen of England as soon as a lot of Irishmen gathered around. Yeah, what a great guy La La turned out to be."

Call them what you will—troublemakers, loudmouths, a real pain sometimes—but without Fitz and La La, a Guinan's or Shannon's becomes as polite and predictable as TGI Friday's—one step from a fern bar. Owners love them, or at the very least tolerate them, because while these guys are holding court, people stick around and drink more—if only to get a word in edgewise.

"I've been no bargain over the years down at Guinan's," Fitz will say, and it sounds like he's bragging. "I almost start fights in here. But I don't feel bad. If someone wants to knock me on my ass, have at it. It's a man's bar." Which is mostly true, except for the "feeling bad" part. Because sometimes he wakes up the next day and thinks he might have drunk too much the night before, and yeah, well, maybe he did go a little overboard with that one fellow. And then he starts feeling kind of melancholy. But that lasts only until he sees the guy again and they get to talking over a beer—though not talking about what happened, mind you, not like girls do about their *feelings.*

"It gets worse if you rehash it," Fitz says.

THAT NIGHT AFTER I first meet Fitz, the moon is just short of full. Kathryn's in California for work, and I'm driving home from an early dinner alone when I find myself passing the turnoff for home and continuing down to the river and Guinan's.

At 6:30 P.M. in the winter, darkness spreads across the Hudson, interrupted every so often by the spotted lights of a tugboat. Except for the neon Michelob and Budweiser signs in the window, and a handful of cars parked outside, it's hard to tell there's any life inside.

I pull in next to a shiny green Jaguar with the license plate ADP INC II and lock my doors out of habit, feeling a little foolish as I peer around at the little houses, which hardly seem menacing. I'm barely through the door when my stomach gives a nervous lurch. Mornings sipping coffee at the bar are one thing. Now it's nighttime, another scene entirely. It's not like Manhattan, where if you don't know anyone, you can sit alone at the bar flipping through the *New Yorker* and no one cares. Here strangers stick out. What if nobody talks to me . . . what if I end up alone in a corner drinking my beer. . . . Come to think of it, I haven't even seen a *New Yorker* since moving out here.

But it's too late because I'm already inside the store and the heads are turning to see who's arrived. There's a nice-looking woman behind the counter—someone I've never seen—who looks to be in her midthirties and is making someone a sandwich. She gives me a curious but not unpleasant look. "Hi there," she calls out.

Hi, I say, I'm just here to have a beer.

"Sure thing, hon, go on back," she says, nodding her head toward the bar. "I'll be in there in a moment."

Squaring my shoulders, I head for the back, wishing I'd just gone home to bed. It's hard to make out the faces gathered around the bar because the only lights are the ones shining from the green and pink stained glass panels around the bar mirror. There are two good-looking guys about my age who are clearly related and both drinking Harp. The younger one is standing behind the bar, poised like a frat boy, while the older of the two, who has graying hair despite his babyish face, sits on the second stool down wearing khakis and a button-down as if he just left the office. I see a much older woman next to him, a thick middle-aged guy leaning over her shoulder, and then—

Fitz, I say automatically, part of me hoping he'll remember me, the other part thinking maybe I'd be better off if he didn't.

"Well, well," he says, his Cheshire cat grin winding across his face. "It's Gwendolyn."

I clearly look surprised when he uses my full name, which makes his grin grow even wider. How did you know my— I start to ask, then realize where we're headed.

"If I told you that—" he begins.

Yeah, I know, I interrupt.

The two brothers are watching, interested. "You two know each other?" asks the older one, his sunglasses still perched atop his perfectly combed hair even though it's been dark for hours.

We spent the morning together, I say.

"Whooooaaa," Fitz calls out, setting down his empty Heineken, and the sturdy guy slaps him on the back. "Now don't go saying things like that where everyone can hear. I gotta wife, you know." Heh-*heh*-heh. That laugh again.

The redness climbs up through my neck and face and I pray no one can tell in the dim lighting. I didn't mean it like that, I mumble, as they laugh with Fitz. Only the older woman doesn't smile, sipping resolutely on her Michelob Light with no comment.

Finally the younger brother behind the bar offers me a beer, which I accept, happy to have something to occupy my hands. He introduces himself as John and sweeps a hand toward his look-alike. "And this is my *older* brother, Ed."

"Yeah, yeah," Ed says, rolling his eyes. I wonder what these two well-heeled young guys are doing hanging out here in the middle of nowhere on a Friday night. Ed holds out a hand, which I shake—Wendy, I tell him—and then I feel Fitz elbowing me.

"Tell them your real name, Gwendolyn," he says. "She's a Tar Heel. Went to the University of North Carolina at Chapel Hill."

I can't help myself. How did you know—— I start again, and my voice trails off as I remember he was in law enforcement and can probably find out most anything he wants from just my license plate. Fitz grins and shifts his weight from foot to foot, clearly delighted he's got me frazzled.

Well, I say, trying to recover, someone's got a lot of free time on his hands.

He likes it. "Oh-ho. See, you gotta watch out for her. With that southern accent." John holds up another Heineken and Fitz nods, but he's not done with me yet. "So, what's with the Gwendolyn?"

She was my godmother, I say. I'm named after her.

"Gwendolyn," he repeats. "The Gwendol—I like that. I think I'll call you the Gwendol."

The woman making sandwiches out front comes into the bar-room, and John leans forward, making room for her to slide in beside him. She's sturdy and curvaceous with blue eyes and a wicked smile that, combined with her figure, make her attractive in a wholesome way you don't see as much in the workout-obsessed city. She is sipping a Harp herself and when she leans forward into the beer cooler, I notice a couple of the guys glanc-ing at the rose tattoo over her right breast.

"That's Jane," Fitz calls. "Jane, meet the Gwendol."

You can call me Wendy, I tell her.

"Whatever you want, hon," she says cheerfully. "Would you like another?" She nods at my Coors Light, which is almost empty. I'm drinking faster than usual out of nervousness and start to shake my head when Fitz says, "What? You're not leaving, are you?"

Well, I was, I say, hesitantly, looking at my watch. Not that I have anywhere else to go. There's just the empty house waiting.

"Well, come on, stay a little longer," Fitz says, the grin again, with teeth.

Okay, just one more, I tell Jane, who smiles knowingly, as if this is a familiar refrain here. She hands me another Coors Light.

So, I ask, are you guys all, like, regulars here?

The older brother, Ed, chuckles. "Yeah, you could say that."

"Tell her how far you came, Donnery," Fitz says to the thick guy beside him, who reaches out to shake my hand firmly.

"Jim Donnery," he says in a booming voice. "And this is my mom, Dorothy." He gestures at the older woman who is sitting, back rounded, hunched over her beer. His mom nods, eyeing me warily.

"I drove four hours from Syracuse to be at the chapel," Donnery tells me.

I nod, thinking it's kind of nice he'd come so far for Sunday church with his mom, but Donnery quickly dispels that notion. "See, some people call this place here the bar," he says. "Others just call it Guinan's. I call it my riverside chapel."

You call a bar a chapel? I ask.

He chuckles. "Yeah, I know. Great, huh? I took it from that Irish song 'Holy Ground,'" he says, leaning further over his mom, who simply curls more deeply over her beer. "There was a seaport section of Ireland called the Holy Ground. The song is about sailors and fishermen who dream of returning from a long journey to the pub, girl and friends they left behind. It's like that in this place when I come back."

This guy Donnery's clearly had a few drinks, but he says his next words soberly.

"I'm pretty blessed to be here."

THE ROOM SPINS with bits and pieces of their stories.

The two brothers, the Preussers, grew up just down the road. They tell me their mother, Nora, runs the local real estate agency, Agnes Donohoe Preusser Inc., and can trace her ancestors back prior to the war. The American Revolutionary War, that is. John works at a golf course and Ed works with his mom. That's his supercharged green Jaguar outside with the business's name, ADP Inc., advertised on its license plate. Their family also owns two of the old Putnam County license plates, 9X-24 and 9X-54, which further denote their historical claim to the area.

I chat for a bit with the eldest, Ed, and learn he has a girlfriend but never brings her here.

Why not? I ask, relaxing a bit. This place is great, I say, gesturing at the old stone fireplace.

Ed smiles at me, as if I'd complimented him. Then he shakes his head. "She wouldn't understand it."

Every so often, the door opens and shuts to the store out front and everyone looks to see who's coming. At some point, a dark-haired guy with narrow eyes and chiseled features walks in carrying his briefcase, clearly having just gotten off the train. "Hi, hon," Jane calls out singsongy, and pulls out a Beck's beer before he can ask. The guy nods and heads straight for the window in the far corner, where he drops his briefcase into a chair and stares out at the river.

"Hey, why are you being so antisocial?" Fitz says, grinning.

The guy smiles over his beer but doesn't look over at Fitz. "I'm getting my bearings here."

"He's a really good guy," Ed whispers, nodding at the dark-haired man. "Drinks here every night."

I watch the guy for a moment, thinking that from a certain angle he looks a little like a count, not that I'm exactly sure what a count is supposed to look like. But the notion, a first image, sticks with me. The Count doesn't introduce himself, just keeps sipping his Beck's alone, looking toward the river as if he were downloading the baggage of his workday into her waters.

The one they call Donnery has come the farthest, swinging by to pick up his mom, who lives across the river. First, though, he eats the sandwich and vanilla milk shake she always has ready for him. "So I'll have a little something in my stomach," he says. "Mom, now, she doesn't much care for beer. She'd rather have champagne or gin. But she'll drink an occasional Michelob Light."

"And it had better be cold," Jane interjects cheerily, reaching down the bar to hand me a coaster.

"That's right," Dorothy chimes in finally in a warm, hoarse growl. I try to catch her eye, but she's still staring straight ahead into the mirror behind the bar, almost protectively, I think, of this space.

"Ma loves the river here," Donnery continues, putting a hand on his mom's back. "When I was little we used to drive down to the bank so we could sit and watch the tankers and freighters go by. And before the Guinans ran this place, Dad as a kid used to row his boat across the river to buy candy and soda here."

Donnery works at the New York State Department of Motor Vehicles upstate, where he says his computer's desktop background is a photo of Guinan's. The day the notice came across his desk proclaiming that license plates could now bear eight letters, he immediately registered "Irishman" for himself. On one hand he wears a wedding ring; on the other a metal band of Celtic knots. "One reminds me I'm married, the other reminds me I'm Irish," he tells me.

Donnery says he makes it down here about every six weeks

and every Saint Patrick's Day. "I keep thinking I gotta keep coming because I don't want to miss the last one."

There's a rustle in back of the bar and Jim Guinan steps in from the kitchen and gives a little wave. "Hello there," he says to everyone. He's supposed to stay off his foot, which is why, I assume, he's not out here at the bar with us. We wave back. My bottle's almost empty, and Jane, through some uncanny sixth sense, is already holding out another with a questioning look on her face. "Another, hon?"

Donnery lowers his voice a bit and leans toward me. "I mean, you know this whole ball of wax folds when Jimmy goes, don't you?"

My stomach gives a funny twitch as I drain the last of my beer. It's 8:00 P.M.

Just one more, I tell Jane.

EVENTUALLY I realize Jane has yet to hand me a bill for my beer. When I ask for one, she laughs.

"There are no bills here, hon," she says, shaking her head.

But how do you keep track of what everyone's drinking? I ask. Looking around, I notice that a lot of the guys just keep laying more money in front of their beers each time they order another. Embarrassed, I dig into my pocket for cash to give her.

"Oh, we've all got our ways. I can mostly keep track in my head just like Mr. Guinan. He can go for hours and then go down the line at the bar and tell everyone exactly how much they owe and what they've drunk. If it gets way too crazy, sometimes I'll stack coasters under the bar to keep count. Another way is to take a book of matches and fold one down for each beer someone drinks, although I prefer the coasters."

Dorothy suddenly gives a big yawn, drops off her stool and walks out of the bar. When she doesn't come back, I ask Donnery if his mother is okay.

"Oh, she's fine," he says. "Just gone out to sleep in the car and wait for me."

But it's freezing out there, I say.

"Oh, don't worry," Donnery assures me. "Mom brings a blanket."

AT 8:30 P.M. I can't understand where the hours have gone. Later, I will come to believe that this place steals time, makes it impossible to be anywhere else punctually, because of some magic grip it holds over those inside.

"You sure you don't need me to drive you home," asks Fitz. Heh-*heh*.

The idea of this is particularly sobering. Nope, I say. I'm fine. I don't have far to go.

As I'm leaving, a pretty, petite woman with long wild hair, a sharp nose and low-rise jeans bursts through the door, grabs a pack of Marlboros from the store and heads back toward the bar. Trotting after her is an overweight blackish hound dog with huge brown eyes who stops to snuffle behind the deli counter for some crumbs. Aside from Dorothy, Jane and me, this woman is the only female to come in here tonight but doesn't seem fazed by this in the least. She brushes past me, dog at her heels, and slips behind the bar to give Jane a hug. Then she reaches into the cooler to pull out a beer, which she clicks open herself. Everyone nods at her.

"Hey, Mary Ellen," Ed calls. She waves and settles down on a stool next to Fitz and drags on her cigarette, completely relaxed.

The dog ambles behind the bar, clearly unhappy with the findings up front. "Don't do it, Lou-Lou," Jane warns the dog. But Lou-Lou starts rubbing her head up and down Jane's shins and looking up at her pleadingly. Before the door closes behind me, I hear Jane laugh: "Okay, okay. I'll get you some cold cuts."

. . . YEAH, I TELL Kathryn on the phone later that night, and there's this guy named Fitz and he's kind of strange, but there's something about him I like. He's a former federal marshal and was in Vietnam. Figured out my name was Gwendolyn, I don't know how. He lives just down the road, in fact. . . .

After we hang up, I lie in bed watching the broad tree branches swing in the wind outside and think about Jane stacking coasters, Dorothy and her blanket and Lou-Lou with the cold cuts and find myself wishing, as I roll over to sleep, that like the woman with the wild hair, I could sweep in there and pull my own beer from that giant red cooler.

THE NORTH CAROLINA seafaring tavern was crowded. My parents were anchored on their stools; the wizened bartender laughed loudly over his own story. At twelve, I was too young to sit at the bar with my mother and father, but old enough that they expected me to entertain myself. It had been this way for hours, long enough for me to read a third of Barbara Taylor Bradford's A Woman of Substance *and consume several baskets of hush puppies in the red vinyl booth nearby.*

Weekend nights at the coast were now a familiar procession of small restaurants and pubs decorated with trawling nets and glassy-eyed trophies nailed to driftwood, their scales lacquered shiny and stiff. I was no longer so young and silly as to be enamored of tire swings and talking birds. Keeping me company instead were my paperbacks. In the worlds between these pages, the female characters were important and escaped from small places to hold glamorous jobs in powerful cities.

My parents ordered another round, and I curled tighter into the booth, wondering how they could sit talking and drinking for so long with the same people. Someday, I imagined, I would live among giant steel and glass towers and become like all my worldly heroines, who sipped red wine from crystal glasses and floated through cities of millions.

the VIPs

Don't you think it's so obvious . . .

After that first Friday night with the parishioners, I wanted in. Wanted to be a part of what was going on down there at Donnery's so-called chapel on the river. Throughout the winter, I dropped by in the mornings to sit with John and in the afternoons to try and catch Margaret in a talkative mood. Friday nights I drove straight to the bar from work in New Jersey, climbed onto one of the five stools in back and sipped the Coors Lights Jane handed me, while waiting for Kathryn to arrive from the city by train.

Here's the thing. Up until this point, I had never really belonged to anything, except maybe my job. No church, no volunteer groups and, forgetting one hapless year of getting my head bashed in boxing, not even a steady sport. Part of it was timing. I'd graduated from the paperback novels of my youth to come of age in the eighties with the glass ceiling already cracked and the notion that a full-time career path was the only one I'd be taking. It was a moment when power suits dominated shopping racks, *housewife* was a dirty word and duty meant exercising all the professional options my mother never had.

Amid the expectations, those unhurried fishing towns and calm ways I knew as a small child suddenly seemed the complete opposite of where I was supposed to be headed. I set my internal compass accordingly in high school and began moving up the obvious first rungs of the proverbial ladder—class president, editor of the newspaper, "most likely to succeed"—before realizing I was even climbing. I had plenty of close friends, but as I sit here writing this, I realize I'm not in contact with a single one today.

My parents were not like me, though they were perfectly accomplished. My father was a veterinarian and mechanical mastermind who could jury-rig almost anything, it seemed, with a little fishing line and duct tape. He could play the drums, ride a motorcycle and navigate by the stars. Like her father, my mother taught Spanish and could catch and clean a fish, not to mention cook it too. They loved their jobs, but learned to disconnect, before the term ever came into vogue. First there was Pettiford and then later a small sailboat they'd use to escape on weekends to remote coves, where no phones could reach and my father's beeper didn't work.

I, on the other hand, began to feel as if a clock were ticking in my brain during my late teens—as if there would never be enough time to finish all the things I should do professionally. In college, my goals crystallized: get to a city, biggest one possible, get hired at a newspaper, again biggest one possible, and put everything else second. I had friends, boyfriends and then, just before graduation, my first girlfriend, a headstrong painter whose cool nonchalance and distaste for regimented anything briefly short-circuited my single-minded focus. Still, our differences doomed us, even as her easy confidence made being with a woman more of a next step than any great leap.

By my midtwenties, I'd more or less sewed up the basics of the life I thought I wanted. Had landed in Manhattan, was writing about fashion for the *Wall Street Journal* and renting an apartment in a perfectly anonymous high-rise right across the street from work. Laundry could be picked up and delivered, food ordered in and anything domestic handled by a staff of building attendants. There was red wine served in deep glasses at swank restaurants where I watched the who's who of fashion and media sup. I sat through interviews with Donna Karan and Ralph Lauren and attended parties where the guest lists included Oprah Winfrey, Cindy Crawford and Madonna. The Internet boom came and went, and while I never became a corporate baroness, I did get promoted to an editor, started earning a bonus and collected stock options. I fell in love again, with Kathryn—a beautiful, blond hard-charger who understood late deadlines and weekend work. And as for those steel-and-glass towers of my novels, all I had to do was look out my living room window and the twin towers soared above me.

The changes, when they came, were subtle. I didn't forget where I came from so much as I forgot things I knew when I was there. Like how to sit quietly and do nothing. If I arrived early at a restaurant, I'd talk on my cell phone or check voice mail until my dinner companion showed up. There were all the gadgets in the world to help me save time, and yet somehow there was never enough time for everything I needed to do. I was always running late, breathlessly hurling excuses at whomever was waiting—"sorry, long interview. . . ." It wasn't the job's fault, nor was there any shame in hard work. Rather it was the *way* I worked, letting that incessant internal clock constantly dictate, that left little time for anything else.

I wasn't oblivious to the evolution. Birthdays passed: twenty-seven, the year my mom gave birth to me; twenty-eight, the year she and my dad bought their first house. Sometimes walking along the river at lunch, I'd think about taking a brief sabbatical just to catch my breath. But like a lot of people, I wondered: Once you're in the professional game, constantly moving toward that next goal, how do you suddenly call a time-out and not get benched altogether? I'd assure myself there'd be time to have it all, just . . . later. Then I'd go back to my desk. Make one more phone call. Answer one more e-mail.

That was the thing about Guinan's. There I watched two grown children working themselves to exhaustion for no other payoff than to keep their parents' legacy alive. Meantime, I had the river and the stately mountains to put my busy little existence in perspective. The pace was unhurried, the visible seasons outside a truer and more sane clock. Inside these walls, I never thought about where else I needed to be, or what else I needed to be doing. And I hadn't felt that way in a long time.

EARLY ON, EVERYONE at the bar was just a two-dimensional snapshot: the moody federal marshal Fitz, the polite older brother Ed, that kind old-timer Mike. As for myself, well, I guess I was initially just that girl. The blonde. The writer. Or whatever else they called me when I wasn't around. Aside from Jane, Mary Ellen and Dorothy, I was in these days sometimes the only woman who showed up regularly. At first my presence inspired awkward silences or mumbled apologies when someone told an off-color joke. At which point a guy might give me a look—like, had enough yet?

But I hadn't. And so I kept showing up.

Week after week, I'd see the same faces. Some, like the dark-haired guy who resembled a count, remained snapshots to me, a testament to the powerful vacuum quality of a bar like this. That is, unless you chose otherwise, your world with the other patrons began and ended at the door—which was where you could coat-check the rest of your life. And so there would always be those patrons I knew primarily by their initial flat descriptors: *the gentle bearded man with the purse, the earnest, talkative fellow on a motorcycle, the guy with a warm laugh and one squinty eye.*

That wasn't the case with everyone, however. As the tiny place slowly made room for me, Old Mike remembered my name, and Jane eventually greeted me with hugs. Ed, the eldest of the two Preusser brothers, began finding room for me next to him at the bar. And Fitz kept calling me the Gwendol and elbowing me when he thought I wasn't paying proper attention to his stories.

So it was that with these endorsements, with each round of "just one more," I faded a bit more into the woodwork. Until finally, instead of raised eyebrows or some nervous guy offering me his stool when I walked in, I got nods and the conversation continued uninterrupted.

And that's when the snapshots started to come alive.

IN WILLIAM FITZGERALD'S case, there were certain facts widely known because he offered them up as part of the steady diet of Fitz: For instance, that he went to Vietnam as an Airborne Ranger in the U.S. Army and later served with the U.S. Marshals for more than two decades. That he loves his dogs, Buck and Ranger, and like Ed is a history buff who thinks Benedict Arnold might have gotten a raw deal from his countrymen, and if you don't believe him he'll be glad to lend you some books.

Occasionally, though, when the beers and conversation flow in sync, there are other details that emerge about the marshal. Such as the fact that Fitz's immune system is pretty much shot to hell from a couple of different illnesses, which he may or may not have picked up from all the crap he breathed in Vietnam. The doctors aren't sure, and hey, who can prove anything, right? And how he can't work full-time because of it all, and misses the high-octane level of the old days when he guarded U.S. Supreme Court Justices in the 1980s (Harry Blackmun and Thurgood Marshall were among his favorites) and supervised courthouse security for high-risk trials such as the Sikh terrorists and the Marxist

Puerto Rican nationalists, Los Macheteros. Of all his duties, though, it was the last assignment he thinks about most, when he served as commanding officer of an elite task force tracking down high-level violent drug fugitives in the New York and New Jersey area. It lasted eight years, ending in 2000—when his health finally became too poor for him to keep working.

"What, you thought I was just a paper pusher?" he says when someone seems surprised at his past. "Why do you think my phone number is unlisted? Look at *Time* magazine, November 14, 1988, page 24. See the guy protecting Imelda Marcos as she heads to court? The balding guy on the right? Yeeaaah, who do you think that is?"

Elbow.

"Listen, did you know that my dog Ranger is dying of cancer? He smells bad, my wife and kids want him gone, but I can't put him down yet; he's my buddy. I'll take another Heineken. Yes, Jane, I switched from Beck's—what, do I need a permission slip? Hey, now, listen to me." *Elbow.* "That stuff people say about me opening the kids' mail and having father-son boxing day with my kid Billy in the front yard, that's about discipline. Where's the discipline these days? . . . What? No, I never mowed the lawn with my gun clipped to my hip. . . . Your neighbors told you that? Well, even if I did, so what? I'd probably just gotten off from work, and besides, what the hell were those people doing watching me cut grass anyway?"

And finally, there are a few bits of information that get whispered now and then but that a guy like Fitz won't always confirm or deny. These are the things that instinctively make him able to command a room and let others know that when he's shooting off his mouth, he's far more than just a loudmouth—that if you scratch beneath the surface, you'll find in this man that there is something to back up his tough talk. That, as they say, there is some real there, there.

Things like what happened on the day he got shot up in Vietnam in 1970 while his six-man Ranger squad was scouting out an enemy bunker forty-five minutes west of Saigon—he can't remember exactly which day anymore: March 20 or March 22? How after he was hit three times—knee, hip, groin—he stayed conscious and returned fire against the enemy for hours until a medevac chopper got him out and flew him to a hospital. And how he was still conscious and saw guys from his company waving at him, and damned if it wasn't one of the best sights he'd ever seen. And that because he had signed a "no notification" paper before going overseas—didn't want his folks worrying over a few shrapnel wounds—his parents had no clue of their son's injuries, only knew he'd been wounded because a friend of Fitz's told them so in a letter. When he got back to the States, drugged up on morphine, two casts on, he had an infection in his hip, which was gashed open like someone had struck him with an ax and had big wires holding it together. And how he had to track his dad down at work. Called him on one of those phones the nurses rolled over on wheels and plugged into the wall. And heard his father pick up the phone and start yelling at him—yelling because he's been scared shitless about his kid ever since he got that letter—"You son of a bitch, what are you trying to do to me?" And how Fitz will choke up thinking about this. How he never cries unless he thinks about his dad. And that because of what he did that day in Vietnam, holding the enemy for so long and with such valor that he was given an honorable discharge and awarded the Bronze Star with a "V" device for valor and a Purple Heart. He's got more medals than that, including a Silver Star, but he won't talk about it. "Just say I was a Vietnam vet and leave it at that."

These are the things that make William Fitzgerald more than just the moody federal marshal.

AND THEN THERE'S that lawyer.

On any given cold morning, a well-timed glance up the landing's road might reveal a bulky long brown coat moving resolutely toward the store. Inside the coat, barely visible, is Daniel Patrick Donnelly, the quick-tongued attorney who practices out of the old brick hotel up the road.

In addition to practicing aviation law, Dan's other most fulfilling pastime, as far as I can tell, is antagonizing Fitz. It isn't just their politics that clash, Dan being your classic white-haired, tea-drinking, Volvo station wagon liberal—"she's got more than 309,000 miles on her"—and Billy Fitzgerald falling more in the *North American Hunter,* flag in his front yard, wood-paneled Wagoneer camp, or as Dan puts it, "somewhere just to the right of Attila the Hun."

Guinan's, however, is their agreed-upon battlefield, and morning maneuvers might go something like this: Dan tucks a copy of a *New York Times* opinion piece inside Fitz's reserved newspaper— it's a story blasting the hypocrisy of the current Republican administration. Fitz then rallies by leaving a towing notification tucked into Dan's windshield while the Volvo is parked outside Guinan's. The attorney tells Billy to expect a call from his lawyer and pockets the ex-marshal's keys when he isn't looking. And so on.

Separating them more than ideology, perhaps, is their style of engagement. Dan weighs in as the methodical champion of the slow, disarming attack. "Now, Billy, don't you think *it's so obvious* . . ." or "Billy, let me ask you a question" might be typical opening strikes for the attorney, while Fitz prevails as the master of blunt parries: "Don't tell me you're going to ask me a question, just ask me a question. . . . What? Now what kind of an asshole question is that?"

Some people think the two men spar because they are so

different, but really it's because down deep, they are the same. Like Fitz, Dan is wired to want the last word. Wired not to lose or give up. Like Fitz, the attorney thrives on attention—and mostly demands it in harmless ways, such as slipping a banana peel into your newspaper or dumping a handful of leaves into an open sunroof. Sometimes, though, like Fitz, he can go too far—as when he feigns a heart attack to tweak one particularly earnest fellow in the bar. It's just a joke, Dan really does like the guy, but when the fellow pales and rushes to call an ambulance, it doesn't seem quite so funny. Then Dan too wakes up in the morning feeling a little bad, and like Fitz, tones it down for a few days.

But this maneuvering, it's just the show. To understand the performance, the motivation, you have to leave the bar. To go back and see the attorney as a kid. And this kid, he spent most of his childhood separated from his natural parents. They were both immigrants—one Irish, one Scottish—working as attendants at a New York State mental hospital. The story has classic tragic elements: Dad was a drinker, Mom was stoic; they worked twelve-hour days, six days a week, and every one was a struggle. The pay was so low that the parents boarded out Dan and his sister to another family while living themselves in a one-room efficiency at the hospital. Dan saw them on weekends, but things slipped between the cracks. Like when the bombs hit Pearl Harbor, Dan was listening to it on the radio, and Mom and Dad weren't around to tell him everything was going to be okay.

So young Daniel Donnelly had two families yet never really belonged to anyone. Became something of a loner, didn't join clubs, spent afternoons riding his bike to the local airport and watching the planes take off, defying gravity, going where people don't go. He mowed lawns to save up and buy his own airplane, but at 15 to 65 cents a pop, that was a really long-term goal. In the meantime, Dan was just a smart kid with a natural ability to lead and an itchy rebellious streak. One day he made up his own club: junior commandos, and after ambushing a fellow kid in the woods, he tossed him into a brook. That didn't go over well at school—*"Hey, but he was the enemy,"* the budding lawyer objected—and this might have been the way his life played out, with Dan always going one step too far, except a teacher named Marietta Ingerson took him aside. Explained that he was indeed a bright young man, but that there were other ways to use your gifts, and didn't he want to make something of himself? It was the kind of talk a father might give, if that father were around and sober enough to pay attention. Marietta put young Daniel at a proverbial crossroads. Told him that while he couldn't change where he came from, there were ways to choose where he went from there. And for some reason, Daniel listened and chose a new direction.

In quick succession this kid finished high school, attempted the seminary, dropped out of the seminary and enrolled in college, where philosophy was his best subject. Then there was a brief stint in the army: 1955 to 1957, most of it performing counterintelligence over in Germany—a lot of methodical research and intuition, but he was good at it. So back to the States, law school, marriage and then the U.S. attorney's office in New York City, where he got a taste of courtroom litigation as a criminal prosecutor and finally found his niche. *This* was as good as commandos—seeing the enemy, strategizing and feeling the rush of engagement.

He received a glowing letter of commendation from FBI director J. Edgar Hoover, along with the chief's photograph, after helping break a case of theft aboard the ocean liner SS *United States*. At first Dan stuck the letter up on his office wall, was even kind of proud of it. But then media stories surfaced about the FBI's harassment of the actress Jean Seberg for her support of the militant Black Panther Party; Seberg eventually committed suicide after the FBI leaked false reports that she was pregnant with a Panther's baby. Dan had no connection to Seberg other than he'd seen her movies—particularly favored *Breathless*—and he had a soft spot for beautiful women. So he took the letter off his wall and mailed it back to the agency along with Hoover's picture, telling the Bureau to dispose of both as they wished. "I no longer want them," he wrote. Made him kind of sad, but hey, he was never big on organized groups anyway.

So Dan eventually left the government for aviation litigation, first at a big firm, then later on his own. And after a childhood of watching airplanes take off without him, the attorney started to take flying lessons himself.

Here, if you're Dan Donnelly, is where the pieces all come together. You finally have enough money to buy your own airplane. You discover this calm place called Garrison, New York, where you join your first club—it's called Guinan's, and the loose structure and freedom of speech suit your personality. You come there four or five times a day for ten-ounce Pepsis, Fig Newtons and conversation. There's a steady family running things here, and a lot of good one-on-one with this one smart-aleck guy called Billy Fitzgerald. The two of you tear into each other, but remember, that's just the show. You're both from working-class backgrounds, both graduates of Fordham University, both staunch defenders of the people and things you love most. If anyone speaks ill of Fitz behind his back, then you're likely to yank them aside and pull a newspaper clipping from your wallet that features a war hero winning the Silver Star. "Billy," you will say reproachfully, "he has one of these."

Sometimes the commando in you sneaks out—like with the fake heart attack—but hey, it's hard to keep our demons buried all the time. Mostly you use your gifts for better things, like giving free legal advice to local guys who can't afford it. And not long ago, at age sixty-eight, you and your wife, Sheila, take a drive up to an island off the coast of Maine. The two of you walk to the town hall, and ask where your old teacher Marietta Ingerson lived. They point you to the cemetery. And so you go there. Take a picture of her grave. And don't you think *it's so obvious* why you do that?

It's just so obvious.

BOTH THE PREUSSER brothers are my age, instantly warm and affable. But while John has the younger brother's typical carefree demeanor, it is the more serious Ed whom I end up talking with most evenings.

Ed is only thirty-three, yet seems to have no place he'd rather be than hanging out with all these old-timers. He's a strange hybrid for a guy so young. Along with his father, Ed senior, he collects and races vintage cars, including a red 1967 XKE Jaguar and a 1956 black-and-white Austin-Healey. The habit has earned Ed junior a nickname, "The Donald," as in Trump. And with his looks, smooth manners and fancy automobiles, he has all the makings of a player. Yet Ed has no taste for a high-octane life of any sort. He doesn't obsess over where his real estate career will take him or what stocks or nightclubs are hottest. In fact, he never even goes into the city if he can help it, except maybe on his way to a Yankees game.

Instead Ed prefers sleepy Philipstown and the company of this pub's crew, wanting nothing more, it seems, than the cards dealt to him. He already knows, more or less, the trajectory his life will take and exudes a deep-set contentment I've never known. Ed resides in his parents' cottage and knows he will move into the main home when they pass on. He can rattle off the names and legacies of all his ancestors down to the fact that his grand-mother died smoking a Salem menthol cigarette in her own bed. He also knows where he wants his own death to lead: to a burial plot he's picked in advance. "There has to be a place where peo-ple can go and say, 'Here's where he is,' " he tells me.

My initial instinct is to brand such complacency as laziness, except that I soon realize Ed is proficient in ways most people my age aren't these days. Things I have to hire someone to do, he does himself—plumbing, electrical work, roofing or car engine maintenance. One Saturday morning he wakes up after a heavy rainstorm to find water seeping into his basement through cracks in the foundation. He whips up hydraulic cement right then and there— actually has hydraulic cement on hand—and fixes them.

Fitz calls Ed an old soul, and Ed takes it as a compliment. He's an avid reader of history who collects old swords, wants to learn ballroom dancing, the piano and to fence with foils—not exactly your typical modern-day bachelor fare. Ed also wants to be married and to see the next road his life will take in the faces of his children. But he's waiting for someone who understands this way of life, someone who can settle in this small town without feeling like they've settled. Given Garrison's population and the fact that Ed spends every Friday night at Guinan's, the odds aren't exactly in his favor. Yet this bar, his time here, they are not negotiable.

"It's hard to explain," he says one cold night, leaning against the wall with his Harp, face lit up by the flickering fire. "Obviously I don't come here to pick up girls because there usually aren't any. It's not about the food, 'cause there isn't any of that either. It's just this calm place where everybody's telling stories and some guy who's a millionaire stockbroker is sitting next to a poor carpenter—and they're equal. You either love it or you hate it."

He puts his beer down on the bar and glances at Jane, who hands him another.

"The woman I bring to Guinan's," he says, looking me straight in the eyes, "is the woman I'll marry."

⟳

4:30 P.M.

There is something akin to a pH level at Guinan's that rises and falls based on the collective temperament of the current clientele at any given time. For instance, the Preusser brothers are more or less pH neutral—can talk with anyone, handle their alcohol without getting sloppy and always leave good tips. They neither instigate nor mediate; rather, their role in the bar's general health is that they constantly participate.

If Fitz and Dan are on the acidic side, able to suck oxygen from the room with a single well-timed barb, then to complete the metaphor, Old Mike is the balancing alkaline force—the guy who makes you breathe a little easier when he's around. He's the one who's actually been around long enough to remember the days when there were only five commuters on the peak-hour train and the status symbol car at Guinan's was a pickup truck whether you were a bank president or a trash collector.

With his electric blue eyes and gravelly voice, Michael R. Mihalik feels familiar even if you've just met him. When he addresses people, it's usually with their name preceding the question: "Tony, hey." "Paul, how are you?" "Wendy, good to see you." And if he's known you for a fair while, he might start attaching the familiar *y* to your name so that Fitz becomes Fitzy, Cliff shifts to Cliffy and so forth. Physically, Mike is solid—could probably shed ten or twenty pounds for health reasons and quit smoking—but there's a hearty quality to his density that lends a certain calming effect to the bar. His stance is wide, both in legs and arms, which he unconsciously shifts and contracts to let people slide around him and into the pub. Put Mike in his usual spot, and Guinan's feels complete.

Above all, Mike is a man of routine. There's a well-worn groove in the bar top about a thumb's-width wide where he stands, and the running joke is that he simply leaned too long in the same spot. Mike has been married to the same woman, Sue, since he was twenty-two and she still cuts his snow-white hair and picks out his clothes. He worked at AT&T for twenty-three years and has spent the last decade managing West Point's telephone system. Every day Mike shows up at Guinan's somewhere between 4:30 P.M. and 5:30 P.M., drinks his Schaefer beer—always Schaefer in a bottle—and goes home for supper with Sue promptly at six.

The routine MO started early and out of necessity. Mike was born just down the road at Butterfield Hospital in Cold Spring. The year was 1944. Just before he turned ten, his forty-two-year-old father, a local banker and avid sportsman, was playing golf with the town judge when he walked up a steep embankment called Killer Hill and dropped dead of a heart attack. Upon his death, young Michael, who was still learning to catch his dad's fastball without wincing, suddenly became the man of the house. From then on, he cut the grass, took care of the plumbing, electrical work, wood chopping—anything that his mother needed at home.

So Mike's nature molded into that of a planner, a can-do guy. And one of the things you do when your father dies early is you make a will. Mike did this in his thirties, around the same time most of his buddies were just starting to learn what a 401(k) was. Another thing he did early was to buy a grave plot—big one front and center at the historic St. Philip's Episcopal Church. Mike is a Catholic, but that's beside the point. Here was where a plot became available, and the way Mike looked at it was "You just have to stay ahead of the curve."

That curve, literally, was located on a stretch of Route 9-D heading north out of Cold Spring. Mike's own son, Michael W., sometimes called Young Mike for convenience, was a few months shy of his twenty-eighth birthday, possessing the same blue eyes as dad, when he met up with Ed Preusser and some friends at a bar in Cold Spring. After a bit Young Mike was ready to move on and asked Ed to come with him. In one of those split-second fateful decisions, Ed said no even though this was one of his best friends. And so Young Mike headed out with another pal, making a left at the stoplight before driving north.

It was after he'd gotten through the tunnel and past the restaurant that he reached the aforementioned curve going pretty fast. The details of what happened next are rough reconstructions . . . a car spinning out of control, another vehicle, an embankment and then a fire.

The police showed up at Old Mike's house next morning with the news. From then it was only a matter of hours before the door swung open to Guinan's. In stepped Old Mike and they were all there waiting for him back in the bar—Fitz, John Guinan and his dad. Ed Preusser was there too, and he was crying. He'd been out to the accident scene that morning. Couldn't help it. Had to see for himself. And the wheels, there was nothing left. Nothing. Just bits of melted alloy on the edge of the road. So Old Mike, he stood in his usual place at the end of the bar. Strong male hands clasped his shoulders, heads shook, beers were opened. There wasn't a lot of talking. A son dying before his father is so hard, impossible, so *not right,* what could anyone say?

Tragedies don't digest well in a small town. At first there's the initial wave of sympathy, which is comforting, like a safety net. At Young Mike's funeral, so many people came to pay their respects that most had to stand in back of the church; they played a Sarah McLachlan song, "Angel," and for the rest of his life Ed Preusser will choke up if he's driving and hears that on the radio.

But then the weeks pass and if you're a dad who lost his son, you can't buy groceries, rent a video or eat out at a restaurant without someone saying they're sorry. Or casting you a sad, knowing glance. What happened becomes part of the permanent description of you, an addendum banked in the collective memory of your friends and neighbors. *You know Mike, his son died a few years back.* . . . And so moving on becomes nearly impossible unless you move out of town altogether. And Old Mike's not going to do that. This is home.

His refuge is Guinan's, where he stays the course with the guys and they never give him the sad glances. He can keep up business as usual—4:30 P.M. to 5:30 P.M. in his place at the bar's end; bottled Schaefer in hand; home for supper with Sue at six. And these routines, after a while, they become their own comfort.

Mihalik also keeps planning. After his son's death, he buys a large tombstone at Saint Philip's and engraves it with a Celtic cross modeled after one that his wife wears around her neck. He inscribes Young Mike's name and life span: Michael W., Oct. 7, 1969–July 27, 1997. And since there's still some room left, he goes ahead and inscribes his own name and Sue's.

Because he knows you've got to stay ahead of the curve.

M*A*S*H *WAS ON* again. *My grandfather put another log into the wood-burning stove and settled back in the leather armchair in the cool basement of his Chapel Hill home. His hands were very stiff, so he alternated wrapping fingers around a tumbler of bourbon and a blue rubber ball. He wasn't feeling so well these days—didn't get to the old fish camp much anymore—and mostly stayed in his armchair watching TV.*

I watched on the couch beside him. The 4077 unit was still at war and things weren't looking so good. But Hawkeye and everybody were in the canteen making jokes and laughing.

I bet Hawkeye misses his family, I said.

"Yes," my grandfather said, setting down the bourbon and reluctantly picking up the ball. "But those people are his family too."

6

healing

. . . I told them right down the line . . .

Winter 2002.

Our neighbor Walter is particularly dismayed by the lack of house maintenance and budgeting skills we display. He clips coupons religiously and sometimes stops by our house to show us his receipts after a particularly successful expedition. "Look at that," he crows. "Triple coupon and I MADE money on the sponges. That's how you should shop." When I throw out a smoke detector that doesn't work, he plucks it from the trash and salvages the 9-volt battery. "This is still perfectly good," he chastens. Another morning, I glance out the kitchen window and see him rearranging the contents of our garbage bags, stomping on plastic water jugs and flattening cardboard boxes into per-fectly aligned rectangles. I walk up the road to find him peering inside a soda bottle to see if it's been rinsed out. "I need to teach you about trash management," he mutters.

I lived six years in the same apartment in Manhattan and saw my neighbors mostly in the hallway or elevator. I never even knew their last names. When I work from home, Walter stops by unexpectedly to say hello, sometimes bringing soy milk he got for free with a coupon. He reminds us to keep the thermostat down and reprimands me for leaving the garage door open during the day.

You're worried about burglars? I ask.

Walter rolls his eyes dramatically. "Not burglars. RAC-COONS. Animals that will tear open your trash and strew it across the yard." He sighs dramatically. "Plus, do you have ANY idea how much money you're wasting by letting the heat escape? What are you, a friend of Central Hudson's?" he demands. Central Hudson is our electrical company.

Walter's wife, Jos, says I can tell him to go away. But I tell her Walter's not bothering me. And the funny thing is, it's true.

"HEY, I NEED your help with something."

I look up from my paper at the bar and see Fitz standing in front of me.

What's going on? I ask warily. His eyes are watery.

"Do you see anything in my eye? I've got something in my eye, but I can't get it out because I've got a splinter in my finger."

I put down my paper and stand up to reach eye level with him. One illness Fitz has is called scleroderma. One day I looked it up on the Internet and read that it's a chronic autoimmune disease that can cause hardening of the skin from overproduction of connective tissue in the body. Some sites tentatively link the disease to the Agent Orange herbicide used by the U.S. military in Vietnam to eliminate the enemy's vegetative ground cover, but nothing conclusive. Still, the scleroderma explains why Fitz's hands are always so swollen and his fingertips appear whitish, hard and thick.

This particular morning, I peer into the veteran's right eye as he rolls it around, but can't see anything. Wait here, I tell him, and go into kitchen, where Margaret is fixing salmon and rice for her father's dinner. I ask her for some rubbing alcohol out of her dad's supplies. She raises an eyebrow but points me to his stash. I gather the goods, pluck a safety pin out of the pharmacy cabinet and then return to the bar.

Give me your hand, I say to Fitz.

Now he looks at me warily. "A safety pin? Jesus Christ."

My dad's a veterinarian, I assure him, as if surgical skills somehow run in the family.

"Yeah, well, I wish your dad were the one digging around with that thing," he grouses but holds out his hand anyway. I dab alcohol on it and begin to prod at the dark spot in the middle of his pointer finger. "It's the cold," Fitz explains. "I have to wear gloves all the time, or my hands just won't work." I push a little deeper with the safety pin and he doesn't even flinch. "They get really stiff, you know. I don't know how the hell I got this splinter today. Hey, listen now, would you hurry up and get going down there. I don't have all day."

It's out, I tell him smugly, holding up the safety pin and wiping off the speck of wood.

"Really?" He looks it over, surprised. "I didn't feel anything. . . . Hey, well, listen, thanks a lot."

It's fine, I tell him.

"No really, thanks," he says, suddenly earnest. "I mean it."

Look, I got to stick something sharp in you, I say, unaccustomed to his being overtly nice. The pleasure was mine, I tell him.

"Heh, *heh,* heh." He laughs loudly, back to his old self. Then he pats me on the back. "You're a pip," he says. He picks up his paper from the Reserved stack and walks toward the door.

What's a pip? I call after him.

He waves dismissively and calls into the kitchen: "Margaret, you behave yourself."

"Hey, I've learned how not to get caught, Billy," she quips back.

FITZ, I yell at his retreating back, WHAT'S A PIP?

JIM'S OLD WHITE 1979 Volvo 244 DL sits parked outside, waiting for the day when his foot has healed and he can drive again. Everyone calls it the "staff car" even though the license plate says "Guinan 1." Margaret has "Guinan 3" attached to her Chevrolet Tracker, and "Guinan 2" is reserved for John, if he ever wants it.

Jim has fond memories of his days in the staff car. He tells me how he'd be driving back from a bar in Cold Spring when the railroad police would pull him over on busy Main Street. They'd usually chat a little, catch up on gossip, and then Jim would be on his way.

"The officers, they'd do it just to say they got ole Guinan in front of everyone." Jim chuckles.

One night, though, he tells me, a rookie officer pulled him over on another road and didn't recognize Jim behind the wheel. He asked if the Irishman had been drinking, and Jim told him no, that he'd swerved to avoid a pothole. The officer asked him to get out of the car anyway.

"Walk the line," he said to Jim.

Jim walked the line perfectly.

"Touch your nose," the officer said.

Jim touched his nose right on target. Then he pointed his finger at the officer and said, "Would you like me to dance a jig right here for you too? I told you I hadn't been drinking."

So you didn't get a ticket? I ask Jim.

He frowns at me. "Well, of course not, luv. I told HIM."

YEARS AGO, JIM played golf regularly with three other Garrison men—Whitney Travis, Ryan Cuneo and Tip Dain—the "Fearsome Foursome" as they were known, at least to each other. On the days they teed up, Jim would rise at 4 A.M. to tend to the store duties, having made up the stock lists the night before for

Peg to give the vendors. Around 6:30 A.M., the men would pile into one car and drive to a nearby course. Usually they'd compete for pocket change because Jim didn't think it was worth playing without a proper goal. It didn't hurt, of course, that he was a scratch golfer in those days either. He chuckles, remembering how he'd come home, pants heavy with coins, and call to anyone in the bar who might be listening—"I got 'em." *Jingle, jingle.* "Got 'em again."

These days Jim's golf clubs are propped up against his private entrance, dubbed the "quality door" because the family used it only when special company came for dinner. The clubs are dusty because the Irishman can't stand on his feet long enough to play a round right now. But he still likes having them there at the foot of the stairs where he can see them coming down from his bedroom—even give the bag a little pat now and then.

One of the Fearsome Foursome, Whitney, died not long ago. Ryan mostly comes in for his morning paper now. Same with Tip, who sometimes slips back into the living room to say hello. Jim, meantime, mostly passes the hours with his bad foot propped on a striped pink pillow while watching USA Network reruns of *Walker, Texas Ranger,* starring Chuck Norris as an old-fashioned ass-kicking Western hero. Around Jim's neck hangs a gold chain with a tiny golf club bag charm linked to it. And as the cool dark shadows of winter darken his cluttered living room, he talks about spring and the day when he'll pull his clubs from the corner and take again to the greens, their freshly cut blades bright and quite alive.

⌇

MARCH. APRIL.

We decide to get a dog. There's a litter of newborn golden retrievers near the *Journal*'s South Brunswick offices, and once a week I slip onto the highway and drive twenty-five minutes to spend my lunch hour sitting cross-legged on the floor while the puppies waddle across my lap, chew my necklace and tumble around in a puddle of whitish-golden fur and legs.

By five weeks old, the puppies are tiny and still shaky on their feet. I don't know how to possibly choose, so the breeder, a warm British woman named Susan, tells me about each one— when they were born, what their personalities seem to be— clingy, raucous, whiny, playful. There's one puppy smaller than the rest. She has a dark ridge on her nose and wags her tail non-stop against the floor. Susan tells me this one arrived twelve hours after all the others and that such latecomers are often still-born and the mother won't touch them. But instead, when Susan walked in that morning, she found the mother licking this wig-gling pup, cleaning her up to join the rest of the litter.

"She was lucky," Susan says. "I call her the bonus puppy."

JIM IS TELLING a story of some sort—something to do with him once kidnapping the governor—when a loud plane roars overhead, making the roof shake and soda bottles clank on the dusty shelves out in the store. It's only a military plane on its way to the nearby airport, but I flinch at the noise anyway.

Jim notices and gives me a curious look from his armchair. "What?" he calls, with a slight twinkle in his eye. "You think the plane is going to fly in here through the front door?"

He's right, of course. But I can't find the words to explain. To tell him that even though we were among the lucky ones, the ones only next to the towers and not inside them, I still have these dreams sometimes. Of exploding buildings or missiles shot through the old apartment, sometimes about the day of the attacks itself. In these last ones, Kathryn and I usually are running down a flight of stairs to get out of a building. But sometimes we're run-ning up, trying to reach Stoli before the building falls.

It happens again when a bunch of us are in the bar and another aircraft thunders by. I try not to move, but the sound strikes a chord somewhere deep, and I grab the edge of the bar tightly. Jim sees this, throws open the window and starts waving toward the sky. "HELLOOOO. HELLLO," he calls to the plane. "We're all in here."

I watch his small arms flailing about outside the window. He looks over his shoulder at me, his eyes dancing. And I start to laugh.

OUTSIDE, THE WEATHER is raw and damp, late afternoon already giving way to the shortened day's darkness. I'm editing a story at home, and it's almost deadline when a loud boom outside makes the entire house shake. I pick up the telephone, hang it back up, then reach for it again and call next door to Jos.

What was that? I ask.

"They're just doing maneuvers down at Camp Smith," she says. "Don't worry."

Are you sure? I sound ridiculous, and I barely know this woman.

She's completely calm. "Yes, I'm sure. Don't worry. If something happens, Walter and I will come to get you. I promise."

WHERE WERE YOU that day? I ask John.

"On a job with Lew," he says, putting down the butter knife. "We heard about the first plane hitting on the radio. The announcer said it was a commuter plane. Later Lew comes running up from the truck and says, 'Jesus Christ, another plane hit.' We tried to keep working, but I couldn't concentrate so we cut off, and I went down to Guinan's.

"The bar was packed. Commuters, locals, they all came here to watch the TV. I guess nobody wanted to be alone. I didn't even know the towers had fallen until I got to the store. I just stood there drinking Harp and watching the TV and going, 'Holy shit.' The phone rang off the hook all day with everybody checking in on folks they knew who'd gone into the city that morning. Dad kept answering and saying, 'Yes, we've seen her. Sure, luv, I'll tell her you called if she calls back.' One woman from Philipstown lost her husband down there. There were a few commuters whose children were in the city that day. They got them out and brought them to Guinan's. One of the kids who'd been at a day care downtown kept talking about the burning birds she'd seen flying out of the buildings before they collapsed.

"Those were people, of course," John says quietly. Neither of us says anything for a few moments. Then John puts an arm around me. "Hey, kid, I wish it hadn't happened this way, but I'm really glad you and Kathryn are here now."

A few days later I'm cleaning up some files at work and come across an old newspaper advertisement for the first movie from *The Lord of the Rings* trilogy. Stripped across the page is a snippet of dialogue from the film. "I wish none of this had happened," Frodo says to the wizard Gandalf. Answers Gandalf: "So do all who live to see such times, but that is not for them to decide. All we have to decide is what to do with the time that is given us."

SCANNING DOWN THE dictionary page, I see *pinprick, pint-size, pipage* and then above it . . .

> **pip (n):** a contagious disease of fowl, characterized by the secretion of mucus in the throat and the formation of a scab on the tongue.

I roll my eyes. Fitz is so rude. I'm about to shut the red Webster's when I glance at a definition above it.

pip: [Old Slang] a person or thing much admired.

I reread it again and, with a small smile, close the dictionary.

EARLY MAY.

At eight weeks, the puppies are stronger on their legs, surer of step. Late one Wednesday night after work, Kathryn and I pull into the drive of the breeder's log-cabin home. We take turns with the tiny wiggling dogs, holding them, watching them scurry about and trying to whittle down our choice. One by one, we carry them back into their den until finally we are left with just two: a big, rambunctious boy and the quieter little bonus puppy. Kathryn is on deadline with a story about the collapse of the Enron Corporation and crouches on the floor with the yapping dogs while talking to our news desk. The boy is more outgoing and lopes about, crying out whenever he wants attention. He's impossible to ignore, and so after a while, I take the bonus puppy back to her siblings and we sit with the boy, watching his antics. It's nearly midnight, and we need to get on the road.

I guess we should take him, I finally say to Kathryn. He's cute, right?

"Yes," she says, juggling the phone with one ear and stroking him with the other. "He is."

I go back to tell Susan we've made a choice, and stop in to peek at the other pups one last time. While I'm watching them, one makes a break for it, scrambling across the backs of the others to rear up where I'm standing. Placing front paws on the edge of the crate, the little retriever tries to climb over the edge, falls back down and then gets up and tries again. I reach down to stroke the blond head and as I'm rubbing between the two brown eyes, I spot the telltale dark ridge across her nose.

"Are you ready to go?" Kathryn says as I walk back into the kitchen. She turns around, sees me standing there, bonus puppy in my arms, and smiles.

She wants to go home, I say.

ANOTHER INFECTION CREEPS into Jim's foot. As before, he doesn't say anything until one morning when Margaret goes to help him put on his socks and he protests—"You're hurting me."

"How can putting on your socks hurt you?" she says. And then she looks at his left foot. Red, swollen, infected. Again.

The hospital pumps him full of antibiotics and saves him. Again. Then he goes to a nursing home to recover and while lying in bed, starts using his right foot to maneuver around. A blood blister develops on his heel from where he uses it to push himself up in bed. After it pops, doctors have to treat that too. But the Irishman makes the best of it. He delights in giving the nurses hell, for instance. Demands they wear gloves because hospitals are the worst place to get an infection. Instructs them on how to take his blood pressure. Reminds them he is a patient, not a resident.

Back home again, Jim is full of war stories. "When they tested my blood sugar four times a day, I told them right down the line, 'Do you see these fingers? They are mine and I need them.' Christ Almighty." We are sitting in the bar alone together on a Friday afternoon when I'm off. The windows are open, and I hear the waves lapping on the shore. Spring is aching to give way to summer; you can smell it on the river first.

The barman's shoulder is nearly touching mine, and I feel his pent-up energy. "Guinan is tough. Now that got around the hospital," he continues. "A nurse came in at one-thirty in the morning, sneaking in to hook me up for antibiotics. I said, 'Good morning,' and she jumped. I told her, 'Don't ever try to sneak up on me because I'm always awake.' I told all of them this right down the line."

It occurs to me then that it's probably been this way for a long time: Jim's doing the telling.

The phone rings, but Jim's foot is propped up. I hesitate for a moment and then put a hand on his arm and walk around the counter to pick up the receiver.

Guinan's, I answer.

THE PUPPY IS too young to stay at home alone while we're at work, but we don't know anyone well enough to hire as a dog walker. One afternoon Jos pulls me aside. "Wally wants to help out, and to tell you the truth I wouldn't mind getting him out of the house for a little while," she says in her rambling way. "But he's too shy to ask you."

So we approach Walter, who agrees to come over twice a day. He pets the bonus puppy and walks her around on a little leash outside. Afterward he feeds her, pats her back until she burps, and then takes her back outside, speaking softly while the tiny puppy roams the yard, sometimes tripping over his big white tennis shoes.

We try to pay him—insist on paying him. But our frugal neighbor refuses to take a cent.

IT'S LATE IN the afternoon and the TV is on in Jim's living room. On the show—I think it's *Walker*—a little girl gets shot, and her grandmother sits at her hospital bed praying—"Don't you die on Grandma. Don't you leave me now." But the little girl's eyes roll back in her head and doom seems certain, until all of a sudden a ray of light beams through the hospital window and into the bed. Angels appear; a choir sings; and the little girl's eyes roll back into the sockets as she awakens, rips off her venti- lator and tells Grandma—"Everything is going to be all right." The doctors pronounce her cured of all ailments.

The Guinans' cousin Kathleen, who is still trying to get a

permanent visa into the United States, is over for a visit and happens to walk into Jim's living room during this particular scene. "And this crap is what you watch all day?" she quips affectionately to her uncle, shaking her head.

Jim keeps staring at the TV. "They do that sometimes on this show," he says. "Just cure them of everything."

MY PARENTS AND I were in the backyard at home playing croquet when my grandmother called with the news. My father ran inside to pick up, and then called for my mother. Something tight in his voice made her drop her mallet onto the ground and rush into the house. She did not come back.

They had found the cancer much too late, my grandmother told them. It was in my grandfather's pancreas and there would be no operation. He'd be coming home.

The last strong memory I have of my granddaddy, he was standing in his baggy pajamas at the head of the hallway, his skin yellow and face gaunt, but tall as ever. I wanted to go to him, to kiss his cheek, but I was too scared of death. And so I sat there, paralyzed, looking at him from the couch. He made it easy on me, though, and waved from where he was standing, smiling weakly as I waved back.

the rising of the moon

One hundred thousand welcomes.

A Thursday.

A rainstorm is brewing, the downpour still dormant inside thick clouds, but you can feel it coming like heavy hands pressing on your shoulders. The full moon is six days waning, its slimming figure nearly invisible. Without the natural light, the country roads are dark and treacherous. Deer linger in the wings of the woods, poised to dart out around the next sharp curve, their eyes saucer-size in the headlights' glare. As I carefully navigate my way across the bridge onto the river landing, I'm impressed that despite the nasty conditions, the parking lot outside Guinan's is nearly packed. Pulling into one of the few remaining empty spots near Georgia the masseuse, I hurry toward the bar, picking up my pace as the faint strain of fiddles and guitars streams through the front door.

Irish Night is about to begin.

AS WITH MANY good things that start small and get bigger over time, there's some uncertainty as to exactly how this whole night got going. For years, in keeping with Irish pub tradition, folks had felt free to belt out a song at the Guinan bar when the mood struck them. Eventually, as Jim tells it, it got him to thinking, and he decided that a little music might be a good way to get people out of the house for a night during the week. So with the help of some local musicians, he began to host a regular monthly music session.

Jim didn't have the money to advertise and figured nobody was going to remember an actual date each month. But the moon, that was about as constant as you could ask for. So he chose the first Thursday after the full moon, which seemed appropriate given there was even an old Irish rebel song called "Rising of the Moon."

Word spread, and soon musicians were showing up from the surrounding counties, some jumping on the train from New York City with their instruments and making the hour's journey upriver to the little pub. With beer as their only payment, the motley crew packed back in the bar on stools, chairs and the edges of tables to deliver a session of Irish tunes for anyone who popped in. Everything was free because Jim didn't believe in charging folks for a good time. When weather permitted, the whole gang moved outside and people lined the walkway in folding chairs, or draped themselves over the metal railings of the train platform, listening deep into the night.

For the most part, these musicians weren't particularly famous. They made a living well under the public's radar—among them were a salesclerk, a tax auditor and a former corrections officer. But everything was different on this night when they gathered alongside the shadowy Hudson River with their flutes, bodhran drums, tin whistles, guitars and fiddles. On this night, Irish Night, they were stars.

I STUFF MY coat behind the counter atop a stack of old newspapers waiting to be recycled and look around for Jane.

The musicians float between the smoky bar and Jim's kitchen, tuning instruments, moving with an air of importance. Beside Jim's sink, a loose-cheeked older fellow in a blue sweater vest and white baseball cap plays a few bars on the fiddle; another guy with swept-back 1960s-style hair leans against the counter beside him, his face easy, cheeks puffing his tin whistle into a high-pitched A note. There is also a dark-haired woman with severe bangs and a body frame that promises to be more than six feet tall when she stands up straight. She's leaning over a gigantic harp in the bar and mumbling to herself.

An affable guy with a clipped mustache and a guitar walks by. "Hi, Candace," he says, pulling up a chair. "Jack." She nods, plucking at her strings and shaking her head. "Man, this is never going to stay in tune in this weather," she laments and continues running her fingers across the instrument. Finally she shakes her head and walks into the kitchen near where I'm standing and begins to sip from a can of Guinness.

What's wrong with your harp? I ask.

"With the windows open and the rain, it won't stay in tune. It's a very sensitive instrument." She looks me over. "I've never seen you here before. Do you play?"

Grade-school piano lessons flash through my head. No, I assure her quickly, not at all.

"I'm Candace Coates, the Highland Harper," she tells me by way of introduction. "I got hooked up with Jimmy Guinan back in 1996 when I happened to walk in here with my harp on my shoulder. He asked me to play a few pieces for him, and I did. Then he told me about this gig. And I've been coming ever since."

Who's in charge? I say.

"Who's in charge?" she repeats. "Well, nobody's in charge really. We all just get together and join in soon as we feel like we know the song well enough. There are a couple of lead players, like Jack McAndrew, who really helped Jim get this whole thing going—see him over there with the guitar?"

I glance where she's pointing and see the guy with the clipped mustache shaking Jim's hand. The Irishman is seated in his green living room armchair, foot propped up on the striped pillow, greeting the players as they drift in and out to store their instrument cases on his couch. His unfinished dinner of salmon and rice sits beside the chair along with an open nonalcoholic Haake-Beck. Even with all the guests, his television hums low; Walker is saving someone else tonight.

"Anyway," the Highland Harper continues, shifting back and forth on her feet, speech picking up tempo, "the lead guys, they'll get things going, but then someone will say, 'Let's play this,' and then that person ends up being the lead player. First night you might only know one or two Irish tunes. But if you're proficient enough, you pick it up pretty quick. We all learn from each other. If someone at the bar wants to join in singing but doesn't know the words, Jack's got a big book there with all the lyrics—must weigh seven or eight pounds. No sheet music, though. That's a big no-no."

Don't you guys get cramped sitting in there together? I ask, looking around at the swarm of musicians flooding the store.

"Oh, I've played in worse conditions. The lobby of the Holiday Inn, train stations, even on a train once. But you do what you've got to do. Last few years I've been supporting myself with the music. Weddings, cocktail parties, lessons. I get room and board in exchange for housekeeping. It's a barter trade. The people who own the house don't mind if I practice at home, which is good."

Well, if you need the money so much, why would you play for free? I ask her.

Candace blinks at me. "For Jimmy Guinan," she says, as if my question were absurd. "I've only missed three of these. Once when I was sick. Once when my car broke down. And once when

I got married." Pause. "But I'm not married anymore. Husband didn't understand the business. I've got a boyfriend, though. And a rat named Zip. I call him Zip because he zips around the bed. He's a fancy breed known as a hooded rat—white, with a sable hood. If you click up to the site Rats.com there are pictures of what hooded rats look like. Most people don't like rats. However, the domestic rat makes a good pet.

"Excuse me," she says, taking a final swig of the Guinness. "I gotta go get a seat before they're all taken—what's your name, by the way? I'll put you on the list when my new CD comes out. Tell all your friends. The Highland Harper."

AS THE MUSICIANS get under way, I see the woman with the wild hair, Mary Ellen Yannitelli, seated behind the bar proprietarily, wearing a straw hat and blowing smoke rings toward the ceiling. She smiles at me, and I nod back respectfully. Mary Ellen is a force to be reckoned with—one of the few women who are just as at home at the bar as the men. Although married with three children, the thirty-six-year-old is far from matronly. She swims in the fast currents of the Hudson River and throws infamous parties where people come in by bus and sometimes don't leave until just before dawn. Her letters to the editor of the local paper are as notable for their length as they are for their scathing rebukes of inane town proposals. Once, the paper refused to run one of her opuses, so she bought a half-page ad to make her point instead.

Half Irish, one-quarter English and one-quarter American Indian, her spirit suggests a woman who's long known how to take care of herself. Growing up nearby, she had eight siblings and parents who were both teachers. Her father also delivered mail on the weekends and was, in his daughter's words, "the cheapest man alive." Once, she says, her dad wrote to the government and asked if he could use his homemade still to make gasohol—a mixture of gasoline and alcohol derived from plants that was a popular fuel during gas shortages. He was quite disappointed when the answer came back no. To save money, he instead turned off the family's oil burner and got a wood-burning boiler. If the kids wanted a hot shower in the mornings, it meant getting up at 5 A.M. to stoke the fire. To have Levi's instead of Wranglers, they were told to get jobs and buy them. Given the number of bodies in the house, her father worried that using the clothes dryer would unnecessarily jack up electric bills, so as Mary Ellen describes it, "there was a clothesline in the backyard two miles long."

Mary Ellen left home when she was seventeen. The impetus was a fight with her folks just before graduation because she'd worn something of her sister's without asking permission. As punishment, her folks demanded she come right home after the ceremony and skip the after-parties. Mary Ellen weighed her options and then stuffed her belongings in a couple of Hefty bags and left.

She stayed with friends and didn't return home until she was twenty-two. Around that time, she met her future husband, Tony Yannitelli, whose ancestors ran Guinan's as a grocery store in the early 1900s, long before the Irishman ever set foot in America. Tony's rugged and handsome Italian features are hidden behind a mane of thick tousled hair and a rough beard that lend a certain gruff quality to his appearance. He seems formidable, not so much because of his physical size but because he exudes a sense of unbridled strength that comes in large part from confidence in his own brainpower. It is an accepted fact in town that Tony is a genius, especially when it comes to the technicalities of woodworking. The richest homeowners in Garrison trust him for the most complicated and lavish architectural details on their million-dollar compounds. Tony could be half asleep and still compute complex mathematical fractions in his head to render a piece of maple or black walnut into a work of art. As a staunch Republican and voracious reader, Tony also makes a habit of rendering silent the liberals who spout off in the bar, unless they can back up their boasts with relevant material.

None of this fazes Mary Ellen. When she first met Tony, she was immediately attracted to a side of him most people never knew: his gentleness. She loved the way he would lie on the floor whispering baby talk to their first dog—the original Lou-Lou. If his workers screwed up one of his complex jobs, as they often did, Tony might mumble under his breath about the "dumb bunch of monkeys" working for him, but he never yelled. Instead, he threw away any imperfections and started over himself. Tony designed his own long wooden skateboards and would ride them down to Guinan's for coffee in the morning, his sons pedaling behind on their bikes with Lou-Lou on their tails. Where other people saw gruffness, Mary Ellen saw a big heart.

She also saw Guinan's differently than most women. Where some were suspicious of this place that captured their men's time, or intimidated by it, Mary Ellen strode right in and managed to make herself a part of it. "Early on, I'd get a lot of flak for being there. Sometimes I'd sense that someone would say something really sexist just to get me going. So at first I'd get a beer and be the wallflower in the corner. But I didn't want to be the wallflower, so eventually I started sitting at the bar."

Partly, her well-honed chutzpah greased the wheels. But more important, she didn't try to change Guinan's—didn't ask it to serve wine or be less smoky or less loud or be anything more than the male bar it was. Having grown up with brothers, she wasn't unnerved by rough talk. She drank her beer—Harp or Corona—

and could banter with the best of them from her perch on a stool: *"What do you think you are, Mary Ellen, the anchor there?" "No,"* she'd *reply coolly, "I'm the buoy."* She wisely kept herself out of the fray, however, by never speaking ill of anyone in town, no matter how much the guys might goad her.

From this vantage point, Mary Ellen had a window into a side of men many women never saw. She watched them strip down their defenses and become extraordinary in their very ordinariness. "I think men have trouble getting to a certain emotional place with anybody," Mary Ellen says. "But give them a space where they aren't being judged and they aren't trying to impress women, and then they can just be comrades and open up. That's what happens here."

I SIT DOWN across from Mary Ellen and finally catch Jane's eye behind the bar. "Hi, hon," she yells and leans over to crush me in a hug. Without asking, Jane reaches deep into the left cooler and hands me a Coors Light. She's given up trying to improve my taste in beer—unlike John Guinan.

"Piss-water," John scowls whenever I order a Coors. Piss-water is apparently a very popular phrase around here.

Well, I like it, I protest. And besides, the governor drinks Coors Light too, you told me.

"Yeah, well, it's like drinking out of the toilet," John counters, his voice loud and angling for the laugh of the crowd.

You're a big talker, I tell him. I'd like to see you tell the governor that.

"I don't care, it's the truth," he grumps. "Besides, the governor's a good man. He should drink better beer. Dad, tell her what Coors Light tastes like." He looks to his father for support.

"Might as well just put it back up in the horse, luv," Jim tells me solemnly.

And so on.

Jane, as is her way, hands me the Coors Light without a word and quickly gets back to work. The musicians are warmed up now and soon lyrics ring around the little pub in a wild refrain—

And it's no, nay, never,
No, nay, never no more
Will I play the wild rover,
No never no more

—and everyone, or at least it seems like everyone, is joining in to sing the chorus. Jane moves fluidly amid the swaying hands, scooping up bills, popping tops, making change and managing to stay on the periphery of the scene even as she's right in the middle of it.

Despite her outwardly cheery disposition, there is an air of quiet disappointment that lingers about Jane just out of reach. Sometimes, when the bar is running smoothly and all the customers are wrapped up in their own world of beer and bluster, I'll see her standing off to the side, absently listening and staring out into the dark toward the hills across the river, toward her hometown.

I went to an alehouse I used to frequent
And I told the landlady my money was spent

Sections of her story, I know by now. She was born in the middle of a snowstorm in 1967, and if you believe in signs, which Jane does, the tumultuous weather was simply a forecast of things to come. As she tells it, first there was her hard-drinking father, who left a few days after she was born. With money scarce, Jane started working at age eleven and by now has clocked time at most of the restaurants and bars across the river.

It was behind the counter of one of them, in fact, that she met her now ex-husband, who was on a lunch break from his construction job. The engagement period was short; one day he picked her up from work, they spent the lunch hour buying rings and the wedding took place that afternoon in Jane's mother's living room. The menu consisted of cold cuts and Pepperidge Farm pastries, and the town supervisor officiated. Jane doesn't have a single photograph to remember the event, which if you are still tracking these kinds of things, was perhaps another sign of things to come. After the couple split up, the children—there were three by that point—went to live with Jane. She now works in a medical billing desk job.

Jane first met "the golden boy," as the Guinan children call their brother Jimmy, in 1998. The elder Jim was already having foot problems then, and Jane had recently started lending a hand at Guinan's to help the family out. Of Jim's two sons, it is Jimmy who inherited more of his father's devilish, freewheeling charisma and personality. During the family's first trip back to Ireland in 1967, they visited the Guinness St. James's Gate Brewery in Dublin where Arthur Guinness had set up the business in 1759. At the end of the tour, every member of the Guinan family got a pint to taste. John, Margaret and Christine pushed theirs aside, but not young Jimmy. He knocked his back, polished off his siblings' and then told the brewery staff to bring more. "Guinness wasn't an acquired taste for me," he says flatly.

I asked her for credit, she answered me nay,
Such custom as yours I could have every day

Jimmy's personality led him to take chances where the more responsible, older John did not. In high school, both brothers made a pact not to cut their hair. John eventually gave in to the pressure of convention, but not Jimmy. "Man, you gotta push the envelope," he admonished when his brother showed up clipped and shorn.

Not content to stick around Garrison, in December 1976 the golden boy struck out for California, found work remodeling homes and thus became the first Guinan kid to leave the family fold. His visits back were usually impromptu and without a firm plan. ("How long you staying this time, Jimmy?" someone would ask. "Till I leave," he'd retort.)

When he and Jane first crossed paths, Jimmy had been gone from Garrison for nearly two decades, long enough to have acquired the air of mystery bestowed on those who leave small towns and don't come back. He walked through the screen door for an unannounced visit that October, the start of the lonely season up north, and met Jane. Pretty soon the two were hanging out together at the store and across the river. Things heated up quickly, and in July of 1999, Jane gave birth to a fourth son. She named him James, after his father and grandfather, James Guinan.

Jimmy had returned to his life in California by this time, and was sending back money for James, visiting from time to time. Meantime, Jane kept helping the Guinans. This wasn't *Leave It to Beaver,* but it was as close to a real family as Jane had known. In the summers, her son James could nap beside Mr. Guinan on the couch and help hold the ice cream cooler open for customers. Jane learned the people's names, remembered their birthdays and what beer they favored. She could even add in her head, just like a Guinan.

And it's no, nay, never,
No, nay, never no more

She also pieced together her own extended family from the customers. Cadets from West Point came to lean on her for a home-cooked spaghetti meal or a place to store a contraband motorcycle off base. She made salads for the poorer clientele who looked undernourished and chicken wings on Friday nights so the guys would have something in their stomachs. If someone was sick, Jane delivered their paper to their doorstep. She also collected stray animals like she did people, as if amassing enough chaos in her life might somehow distract her from how life hadn't exactly gone according to plan.

> *Will I play the wild rover,*
> *No never no more*

As for her own father, Jane says she never heard from him until 1995, after she tracked down his address and wrote a letter promising that whatever choices he'd made, she would always be there for him. A few weeks later, the phone rang. "Jane, is that you?" came the voice on the other end. "It's your father. I'm at a pay phone." They talked briefly. He promised to come by for Christmas but never showed up.

Then in February of 1999, Jane says she got a phone call from a woman she'd never met. It was the granddaughter of a friend of Jane's father. He was in a veterans' hospital in Manhattan, the woman said. He'd been mugged, left for dead and in and out of a coma for several months. Jane went to Manhattan to see him and found the spitting image of her lying in the bed there. She stayed for three hours and says her father said he was sorry for everything he had missed out on. When Jane left, she gave the nurses' station her phone number and asked them to call her if something happened to him.

> *I'll go home to my parents, confess what I've done,*
> *And I'll ask them to pardon their prodigal son*

She says she never heard from them or her father again.

THE MUSIC DROWNS out the rain, which the skies have finally unleashed across the Highlands. I notice the Highland Harper has given up on her instrument and resigned herself to strumming somewhat morosely on a bodhran drum instead. I'm standing near the end of the bar in Old Mike's usual spot. My Coors Light is nearly empty, which Jane notices as she's passing by me into the store, a box of empties hoisted up on one shoulder. A line of customers is waiting out front to purchase cigarettes and sodas.

"Ready for another, hon?" she hollers over the din.

I nod. No hurry, though, I tell her.

"Well, if you don't mind, just reach in there and grab it for yourself while I take care of these folks out here."

And there it is: the invisible red velvet rope lifting. Or Guinan's equivalent of it. I half want to go outside and make a call, tell somebody, except, well, who's going to understand this? So I slide carefully behind the counter. Metal handle is cold under my hand, smooth, as I lift the cooler top—and it's heavier than I thought. Can't see inside, don't have it memorized like Jane does, so first beer out is a Budweiser. Slip it back quickly. Reach farther in, pull out another bottle and hold it up to the light. Coors Light. Click off the top. Then I toss money into the wooden box and return to Mike's spot at the end of the bar, the crowd sort of swelling and contracting to give me room as they sing along.

Maybe it's the alcohol, but the music seems to lift up the tiny wooden space until we kind of levitate above the river in a universe all our own. I don't know the words and can't begin to carry a tune, but suddenly I don't care anymore. So on the final chorus I join in:

> *And it's no, nay, never,*
> *No, nay, never no more*

AROUND 11 P.M., the bar grows quiet. I'm out in the store putting on my coat to leave when Mary Ellen comes dashing around the corner, shushing people. "Come here, come here." She waves me into Jim's kitchen. "This is really special."

I follow her in. Jim is standing by the door to the back stairs of the bar, facing the crowd, his back to us. He's wearing khakis, his yellow golf polo with his initials JG on the pocket, white hair slicked back. I wonder if he's going to give a toast or a speech of some sort. Nobody in the bar is moving. The musicians have all set down their instruments and are watching the Irishman attentively. The only sound is the creak of the sailboats' lines rubbing against the pier outside.

After a pause, he begins to sing.

> Oh Danny boy, the pipes, the pipes are calling
> From glen to glen, and down the mountain side

His voice is surprisingly clear, even a cappella. He holds the side of a chair, leaning forward slightly, his soft, sweet lilt filling the room, which is silent. Jane doesn't open a single beer. No one shifts on the stools. Not a soul speaks. I have to look twice in the darkened bar to make sure, but it's true—over in the corner, two women are actually crying.

> And if you come, when all the flowers are dying
> And I am dead, as dead I well may be
> You'll come and find the place where I am lying
> And kneel and say an "Ave" there for me.

When the Irishman finishes, his voice kind of drifts off out the open window. For a moment, there's only the soft patter of rain on the roof. And then a heave of applause, and they are upon him, shaking his hand, introducing their girlfriends—"Honey, this is Jim Guinan. The one I've been telling you about." "Well done, Jimmy." "Good as ever."

I stand back in the kitchen with Mary Ellen away from the fray until he passes by us.

"Hello, luv," he says cheerily. "Are you enjoying yourself?"

Yes, I say, suddenly feeling like an outsider again. I didn't know you could sing, I tell him, plucking at the wet label on my bottle.

"What, you've never heard me before?" he says, surprised. I shake my head. "Well, come here, dear." And he takes my hand, pulls me down toward him and kisses me on the cheek. *"Cead Mile Failte,"* he says.

I look at him, confused.

"It's a Gaelic greeting, luv. It means 'One hundred thousand welcomes.' "

———

At my grandfather's funeral there were so many people. People I'd never seen before—students, janitors, university officials, local storekeepers—taking up all the plastic folding chairs outside in the cemetery until there were none left and people stood instead.

I sat in the front row between my dad and aunt, squinting in the sun at all the foreign faces, searching them for some trace of my grandfather. They stared back, shook their heads solemnly and nodded. This was my first funeral, and I didn't know what to do. So I nodded back, wondering why I didn't feel as sad as they seemed to.

Then people started talking. They told stories, ones I'd never heard. Everyone had something different to say about how William Albert McKnight was such a great man, an academic, a friend, a mentor. And I began to feel strange, like maybe I should say something, tell them about the Cadbury Cream egg in the jar because I bet they didn't know about that. And thinking about that suddenly made me cry. And I cried so hard my chair shook and my nose ran down onto my dress because I didn't have any tissue, and suddenly everything was over and people were patting me on the back, clucking their tongues and telling me it would be okay.

I tried to stop crying so I could talk. But everyone was getting into cars, and so I never got to tell them about the egg and the jars with skate egg purses and the SS Wendy B., *and all those things that made this great man of theirs my grandfather.*

the journey to god's country

You make a better life for them than you had.

They called that World's End," Jim says to me, pointing to the spot off West Point where the Hudson narrows to make a sharp S-curve.

It's a bend whose treacherous currents and crosswinds had conspired more than once to send early explorers and colonists to a watery grave. However, for George Washington and his soldiers, Jim points out, World's End was a strategic boon. Washington knew the British would attempt to seize the Hudson, thus shutting off contact between the northern and southern colonies. The mountains and the river's twists and turns here were the greatest natural defenses available. To deter the enemy, a chain was forged—with each link stretching two feet in length and weighing 180 pounds—and then floated across the river at the sharp turn.

"And a fine idea that was," Jim says.

Isn't it true, though, I ask, that there was an earlier chain that broke—and this second one was never even tested by the British?

The Irishman stares at me for a moment. "Well, that's not the point, now is it, luv? The point is that they kept on trying."

JIM WAS BORN on December 27, 1925, into a country without widespread electrification or mechanization, and where poverty was the crushing sort of Frank McCourt's *Angela's Ashes*. Even when technological advances such as running water, working stoves and electrical lighting were creeping into Ireland's capital of Dublin, rural towns often remained without. In Jim's hometown of Birr, which sits in the middle of County Offaly, shops were mostly mom-and-pop operations where customers short on cash could take what they needed and add their names to a store's "good book." In those days, children went to work as soon as their families needed them to. Women ran the households, handling the small livestock, washing, cooking, managing the children and their affairs—all this without any of today's modern conveniences.

So it was with the Guinan clan. Jim's father, John Guinan, was a highway ganger who supervised a crew of men constructing and repairing the roads of Birr. His mother, Kathleen Clancy, ran the household of thirteen children, cooking for them over a fireplace with a hob grate, scrubbing all their clothes on a washboard and pressing each shirt with a heavy iron heated over flames. The children bathed with water carried in from the outside by pail and warmed over the stove. Lighting came from paraffin lamps, which smelled better than kerosene. Jim's mother was the boss, and if dinner was at five then it was at five and you had better be there, hands washed, shirt tucked in.

The family had three bedrooms for the thirteen kids and their two parents. The boys slept in one; girls in another, all of them packed horizontally in the beds to fit more bodies. Each bedroom had a fireplace, which in the winter burned constantly with a mix of wood and peat turf cut from the county's plentiful bogs. That kept things plenty warm until about 2 A.M. or so, when the embers began dying out. Then the children would press close together and pull their heads under the blanket until their mother came in to rekindle the flames, her breath powdery and quick in the cold air.

Still, it is the natural richness of his homeland, not the lack of material goods, that Jim remembers so vividly. County Offaly sits in the middle of the country and is comprised of long, low swaths of grass that rise to meet the rugged Slieve Bloom Mountains. Marking the territory is the River Shannon and the great tracts of peatlands that are milled to produce what's known as the "brown gold of Offaly." One of the main attractions is Birr Castle, a sprawling stone structure with turrets, terraces, a suspension bridge and a giant telescope where the Earls of Rosse have lived for nearly four centuries.

The Guinan home was situated next to Birr's golf links, and from age five, young Jim was swinging his own clubs, hand-me-downs from the nearby club. He was patient and taught himself on the wooden shafts, hitting thousands of balls out into the field beside his house. Before he was ten years old, he could drive a ball "oh, about 150 yards, I'd reckon," he tells me. Better yet were his putting skills, which he honed patiently whenever he could sneak onto the nearby greens. It was a good way to get out of the house, which was always frantic with the comings and goings of his siblings.

The main center of traffic was the kitchen. Eggs and fresh milk were delivered daily from local farms. No child could leave in the morning without a full stomach, because school was a three-mile trek, which could take upward of an hour in the winter. On these mornings, Jim and his brothers and sisters would walk shoulder to shoulder in their coats, ambling along, cracking jokes and picking up friends along the way. The colder it was, the faster the pack moved toward the tiny stucco frame schoolhouse, where another fire was lit and their teachers waited.

With his quick wit and fast tongue, Jim excelled at his early studies. But like so many others his age, his education was cut short when his father's roadwork didn't bring in enough to meet the family's needs. At age fourteen, Jim ended his academic studies and went to work in a shoe factory. He was in eighth grade. From that moment on, he would work almost every day of his life for the next sixty years.

For his first job, Jim rode his bicycle five miles to the factory. By 8 A.M., when classes back at school were just getting under way, he was pounding out heels with a little hammer. That was a job for novices, though, and after a time, he graduated to toes, which involved heating the pointed part over a flame and passing it to the next man, who would add rivets and then send it along for stitching. After eight hours hunched over his work, breaking only for lunch, Jim would mount his bicycle and pedal back home for supper.

The shoe factory didn't pay well, and he soon left to become a butcher's apprentice, a far more lucrative profession. With his boss, Jim would roam out into the countryside, helping pull in the cattle for slaughter. Part of young Guinan's job was to tie a heavy rope around each animal's neck, fasten it to a sturdy ring on the floor and then stun the creature with a quick blow of a hammer to its skull. Once the cow fell to its knees, Jim was quick in with the knife to cut out the heart, using his foot as a pump on the beast's bowed back, an action which helped push out the blood into a bowl for pudding. This way, there was no waste. Then Jim would wind the cow up, back feet first, head down and skin it.

It was gruesome work for a boy. But there was no room to think about such things. "Didn't matter," he says now matter-of-factly. " 'Twas a part of me job."

IN 1950 HE met Margaret Curnyn, better known by the Irish nickname of Peg. By this time, twenty-five-year-old Jim was working over in England for the Birmingham Small Arms Company. One day he noticed a tall, pretty girl with jet-black hair working down the line from him. She was strong, the daughter of farmers and of the same background as Jim, which is to say that if there was no one to help you, then you helped yourself. Peg had left school in the sixth grade to start working, and her career path had ultimately taken her to the same arms plant as Jim. In these days, the Irishman looked a bit like a young James Cagney with his slicked-back hair, arched eyebrows and mischievous half smile. Soon Jim began concocting excuses to walk by Peg, giving a quick nod and grin every so often as he did, and amid the hot clatter of the machines, a romance kindled. On September 19, 1951, they were married, with John, Jimmy, Margaret and Christine coming in quick succession in 1952, '53, '54 and '55.

To leave Ireland was never something Jim wanted. It was his home, even when he was living in England. The open space, the peat, the scents that changed with the seasons—wildflowers and mowed hay—it was nothing to give up lightly. But with more mouths to feed, and the Irish economy still struggling, Peg was insistent they try their lot in America, where members of her family were making a go of it. And after his fourth child, Christine, was born, Jim agreed. "You make a better life for them than you had," he says of his children. "That's the whole idea of bringing up a family."

Coming across the ocean in the *Queen Mary,* he kept any regrets he harbored well masked from his children and wife. That first day in the Bronx, amid the noise and cramped quarters, he began to wonder—was this all there was to America?

And then they arrived in Garrison.

As in Birr, Jim suddenly found himself amid open plains and valleys, a strong river and lush rolling hills that spired upward into small mountains. This place too was a spot with mom-and-pop stores—three of them on the hamlet's river landing, where they rented half of a small house. Just a short walk up the road was a lively bar called Bosco's Folly where the owner served up enormous quarter-pound burgers, frying them one by one in a cast-iron pan. There was a golf course, and as Jim liked to say, even a turreted castle way up on the hill, for God's sakes. Why, if you threw a log of peat in the fireplace and looked out the window as the snow quieted the land, a man might imagine himself back in Ireland.

Two days after they arrived in Garrison, Jim went looking for work, walking several miles to help build a roof at a farm owned by a wealthy town businessman. After all his days pounding out heels and toes in the factory, the Irishman only owned a single pair of leather shoes, and kept slipping and sliding precariously on the rooftop. The next day a lady saw him walking back to the job and offered him a ride, which seemed to Jim a good sign that folks trusted each other around here. As soon as he got his first paycheck, he bought himself a new pair of shoes. After a time, he saved up and bought a 1950 black Studebaker for a few hundred dollars and parked it right outside the front door.

There was no money for toys, but the kids learned to entertain themselves outdoors, building homemade go-carts and whipping around the hairpin turns of the empty mountainous roads until they hit a fence, toppled over and picked themselves up laughing to race back uphill and do it all over again. There was fishing, woods to build forts in and faithful dogs to walk them to school and back. The gravest dangers were the railroad and the river, "the devil and the deep blue sea," as Jim christened them. He taught the kids to stay away from the tracks, and after his youngest, Christine, fell in the Hudson a couple of times, he made sure they all learned how to swim.

He also encouraged his children to be tough in other ways. In the early years, children of the richer families often scoffed at the Guinan children's strange brogue, handed-down clothing and leather-lattice sandals. "The penny millionaires," Jimmy dubbed the rich kids. "They used to tell us they flushed their toilets so we'd have drinking water." The teasing got so bad that John at one point decided he wanted to go home to Ireland. So he packed up his little suitcase with pajamas, took a quarter and walked over to the train station. When Jim discovered his son huddled up there, he came and sat down next to him.

"Where are you going, lad?" he asked calmly.

John told him.

"I see," his father said. "Well, that's a long way, as you probably know. You'll need more than twenty-five cents." And he reached in a pocket and gave him another quarter and left him there to think things over. After a bit, when no one tried to talk him out of his mission, John abandoned his quest and came inside for supper.

As often as he could, Jim wrote to his mom and siblings back in Ireland to reassure them that he hadn't lost his mind by crossing the pond. "Ah, yes, things are quite good. Really they are. You see, I've landed me'self in God's country. A man could die happy here."

⁓

THE ADULTS TALKED about what to do with my grandfather's office. He had worked in the same basement nook at the University of North Carolina for as long as I could remember. Number 134 Dey Hall. The room was cluttered with papers, boxes and Spanish texts piled high to the ceiling. The only luxuries were a couple of chairs for students and a manual typewriter he would hunt-and-peck his correspondence on. When he was alive, students often let his dog, Quesa, into the building, where she would wander its corridors listening for her owner's deep voice. Upon locating him, the collie would sidle into his classroom and lie by his feet until the lecture was finished, after which they would go down to Number 134 together and while away the afternoon, my grandfather hunting-and-pecking, Quesa asleep with her belly pressed against the cool floor.

Even after my grandfather retired, he never stopped going to the office; he was comfortable there with his routines and familiar mess. There were rumbles that the other faculty wanted the space, wanted to turn it over to new, younger professors, but my grandfather was such a presence that no one could quite bring themselves to make him leave.

He kept Number 134 until his death.

9

the bookman

There's talk, you know. . . .

The rumor first reaches me inside the Australian book dealer's shop.

David Lilburne's office on Garrison's Landing, which he runs with his wife, Cathy, is a rich niche of antiquarian books, old maps and rare etchings. His brindle boxer, Winston, sleeps on the couch, occasionally chewing on the corner of documents, and unfortunately, he is not discerning about their origins. David doesn't fret about appearance—would rather wax eloquent about the smell, feel and leather of a particular binding than spend time organizing—and so his store, Antipodeon Books, has acquired a rather warm, mishmash feel. Stacks of books maintain a sketchy semblance of order: the shelves inside are grouped into sections—Antarctica; Hudson River; Travel. There is a smidgeon of fiction, including a first edition John le Carré, *The Spy Who Came In from the Cold,* inside of which the author has sketched his own face by way of an autograph. The various tomes drift in literary vines onto the porch, where in the summertime, they tend to huddle in complete and fantastic disarray.

David is slim, tall enough to reach his top shelves without a ladder and always in motion. When he speaks, his facial expressions offer context to his words—raised eyebrows, wide grin, conspiratorial tilt of the head. He has been a natural salesman since his twenties, when he stood in London's Victoria Station hustling travelers to stay at the local youth hostel where he worked. It was there he met Cathy, a beautiful, headstrong young American who'd grown up near Garrison and was studying at the Sorbonne in Paris. David approached Cathy and her three friends as they stepped off the train, offering his hostel's accommodations for the night. One girl said yes, another said no, Cathy said maybe, which seemed like good enough odds to David. So when the girls told him they first needed to check on accommodations at another location that were cheaper, David sealed the deal when he asked, smoothly—

"Do you have any English change for the phone call? . . . No—well, then, here, take some of mine."

Fate cooperated with David and the alternate location wasn't available. And the rest, well, that became their history.

Among other traits, the couple's intellect and gypsy tendencies made them a comfortable match. One summer before settling down, they bought a little Toyota Hilux pickup truck with a camper and drove across America, sleeping in back because they couldn't afford hotels. Likewise, after getting into the book business back in London, they would return each year to the States and roam about the country in a station wagon that redeemed its gas-guzzling habit by fitting a fair many books in its fat back—as well as David and Cathy if no accommodations could be found, or afforded, on a particular night.

Now they were parked, more or less, in Garrison, where they'd amassed a formidable collection of materials from the world's greatest explorers, as well as Hudson River artifacts, including old ferry schedules, maps and drawings. From their perch on the landing they could watch the river itself easing by outside their window.

"There's talk, you know. . . ." is how David puts it to me one Saturday afternoon. I'm flipping through etchings and aquatints of flowers and fish, trying to find a last-minute Mother's Day gift. Winston is draped across my lap, snoring loudly, his cheeks puffing out with the exertion.

I glance at David quizzically and wait. The bookman is often disconcertingly vague.

Talk of what? I ask finally.

He wiggles his eyebrows. "Talk about what happens if Guinan's closes."

What do you mean? I say, staring at him.

David narrows his eyes and leans in toward me, his Aussie accent thickening. "Well, it's not like John and Margaret can keep this up forever. And that's prime riverfront property there. Word is some folks around here are interested in turning the place into something more upscale. Maybe a fancy restaurant or something."

Cathy peeks out from her office nook. "Guinan's gone?" she says. "But where would everyone go?"

I imagine some thin-lipped waiter standing in Old Mike's spot, and white-clothed tables replacing the battered green stools . . . there would be long candlesticks lighting up the room instead of the tiny Guinness lights and a maître d' checking coats in Jim's living room. . . .

I shake my head. The whole scenario seems impossible until I hear Donnery's voice echoing in my head—"you know, this whole ball of wax folds when Jimmy goes, don't you?"

And that's when it dawns on me. With Guinan's, we were all living on borrowed time.

It turns out rumors of Guinan's imminent demise had been floating around for a while. In small-town fashion, they would begin with a kernel of truth and then pop into something bigger and more exaggerated as they spun from one heated tongue to another.

The first round of talk surfaced not long after Peg died. *You know, she really ran the show,* folks would whisper. *Typical strong Irish-woman. How do you think old Jimmy will hold it together alone? Well, I heard a few folks in the tenants' association want him out anyway. That's valuable riverfront property.* And so it went, with everyone adding their opinion until they forgot what was opinion and what was fact.

Then for a few years, Jim's youngest daughter, Christine, came to take over the business with her husband, Mike. That was in 1994, and under their command the rumors quieted down for a spell. Over Jim's protests—"There's nothing wrong with how things are done now"—they upgraded the fixtures, added perco-lated coffee, expanded the inventory and nearly doubled the store's business. But then the gas concession down by the dock got removed because of soaring insurance costs, ending a steady source of income and foot traffic. And Christine and Mike grew frustrated battling Jim over their every decision. So they turned the reins back over to him and eventually moved down to Florida to help care for a sick relative of Mike's. And then the rumors started up once more, with folks speculating how long the store could survive without Christine and Mike and the gas conces-sion. And sure enough, everything eventually got reduced to the same breathless end—"Did you hear Guinan's is closing?"

And still Guinan's just kept going.

Problem was, this wasn't some 7-Eleven. People cared so much about the place that they'd get a little crazy from the rumors—start talking about petitions, lawsuits, boycotts—anything to keep

Guinan's afloat. Every time help was short, neighbors and friends pitched in—some of their kids would bartend during summer college vacations. But, of course, the more impassioned people became, the longer the rumors tended to stay around and ferment. And now that Jim was sick, it had all started up again. Which was how David had come to pass along the latest chatter to me.

Anyway, like I said, such talk could make a person do crazy things.

I WAKE BEFORE five. Rolling to my right side, I face the deck doors and for a moment watch the shadows of the strong trees, their branches heavier with spring's arrival.

On the chair lie my work clothes—an expensive Jil Sander suit, which suddenly seems ridiculous, like a floor sample waiting for a naked mannequin, not me. I pull on jeans and a fleece pullover and yank my hair into a ponytail. As an afterthought, I reach into my briefcase, pull out a reporter's notebook and tuck it into my back pocket.

In the short drive, early light reveals the outside world already busy. A herd of deer lope in the grassy field, racing parallel to the car until, suddenly spooked, they veer off into the woods. The breeze plays off the hills of the North Redoubt to my left, sending the scent of the ebbing spring night sneaking through the open sunroof. Turning right toward the river, I wind past the governor's house and onto the river landing.

John is wrapping doughnuts in plastic wrap when I walk in. He looks at me and then back down at his task.

"Morning, kid," he offers. "You're up early."

I'm unsure what to say. Unsure even what I want, except that I want to be here with him. That I want to be in this place as much as I can before it all disappears. The store is warm; the compressors of the ice cream cooler and refrigerators hum methodically. There is butter softening on the counter. I smell coffee. And as always, the familiar sweet scent of sugar mixed with a hint of earth tracked in on hundreds of heels.

Can I take over there? I ask finally.

John looks me in the eye, deciding something. Then he motions toward the counter and a half-wrapped jelly doughnut.

"I do it like this, see."

My parents, aunt and grandmother drove to Pettiford with my grandfather's ashes. All four of them piled into the SS Wendy B., *her aluminum frame now tired and dirty from neglect. Slowly they motored to the mouth of the creek, where they scattered my grandfather's ashes downwind into the river.*

Someone else was living in the old fish camp by then, and they had started to fix up the place, building a deck, pruning back the oak trees. The change upset my mother, even though my grandfather was gone. One day I caught her in the kitchen looking at old pictures of the tire swing and weather-beaten pier.

"This is how I'll always remember it," she told me.

part

II

Rivers know this:

there is no hurry.

We shall get

there some day.

—A. A. MILNE

10

morning

Just like two pianos, upright and grand.

I can't add.

This much is clear the first time a harried businessman in an expensive suit stands before me with an armful of newspapers, coffee, a bagel and cigarettes. Mostly the customers make their own change from the coins stacked on the counter, except when they have big bills, like this man with his twenty. I try to tally his purchases, but every time I get close to the total, I choke and my fragile mental sum collapses into a pile of 5-, 10- and 25-cent digits. The man before me drums his well-manicured fingers on the metal counter, which doesn't help. People are lining up behind him. I hear the train approaching in the distance, and John is busy in the kitchen so he can't save me.

My face flushes as I apologize. I want to defend myself. To assure him that I'm not a complete idiot. To explain how I've used calculators since grade school and graduated into a workforce with smart software and spreadsheets that not only add for me, but also spell and correct lousy grammar. How at the *Wall Street Journal,* we have a whole staff of smart copy editors who will double-check my arithmetic. And that if I get something wrong the first time, we can fix it for the next edition and half the country will never know.

But all he wants is his change so he can get to work. Turning my back to him, I stare inside the wooden cash register as if some mathematical genie might appear, and then in desperation begin using my fingers to count against my leg like a kindergartner. I pray no one speaks to me until I finish . . . $.75 . . . $1.00 . . . $.60 . . . $5.95 . . . $.50 . . . $.50 . . . okay, $9.35 . . . wait, no, $9.30. I stick the twenty-dollar bill into the drawer and pull out $10.70.

Not enough coffee in me this morning, I say humbly, handing the man his coins and bills.

He flashes me a wan smile of pity and hurries out the door.

TRAIN COMMUTING BEGAN in Garrison in 1849 as the Hudson River Railroad extended its reach north through the Hudson Highlands. It was on the eve of a period that would see a spectacular rise of industry in America, as railroad tycoons laid tracks across the young country with tumultuous speed. It was also when the foundation for Jim Guinan's home was put down. The builder was a retired New York City wholesale grocer named Henry White Belcher. This distinguished, white-bearded gentleman had acquired sizable property holdings along Garrison's Landing—including the last parcel, which would one day become Guinan's.

Build next to the trains, Belcher knew, and the customers would come.

JOHN TEACHES ME in between each commuter wave. As it turns out, there's a long-standing way of doing almost everything here.

"Okay, if you leave the handle on the coffee filter pointing outward like this, then some last-minute person in a rush is going to hit it with his hand and spill hot grounds all down his shirt. So turn it to the side like this, see? And you've got to pull apart lids for the coffee cups so they aren't all stuck together because people are already juggling briefcases and umbrellas and whatever else they're lugging around, so they don't have time to fumble for a lid. And you can't let the pots get empty or you're gonna have a stampede of folks knocking into each other waiting for a fresh pot."

I follow him around the store, taking notes as I go—*pull apart lids. . . .*

"What the hell are you writing?"

I'm trying to get all this down so I won't forget, I tell him.

"Jesus Christ," he says. "It's not brain surgery. Okay, never mind, now the governor's newspapers—you've got to set aside two copies of the *New York Times,* two *Daily News,* two *New York Posts* and two of the local *Journal-News.* Stack them neat on the end of the counter and write the name Weiss on the top. One of his security guys will come down and pick them up. And then——"

Who's Weiss? I interrupt.

"Used to be the state trooper who lived around here and worked with the governor. Now, when you get done with that——"

So why do you still put it under Weiss, I butt in again. Why not just put the governor's name, Pataki, at the top?

"I don't know, that's just the way Dad does it." He rubs the top of his balding head impatiently.

And why two? I ask, confused. Why does the governor need two copies?

"Who the hell knows? One of the troopers told me his wife, Libby, got tired of sharing. Now did you have any other questions I can answer for you this morning or can we move on?"

I follow him into the kitchen where he shows me the proper way to slice up old bread for the seagulls' breakfast——Duck

Detail, as Margaret calls it. "See, this is how I do it," John explains. "Small enough so the birds can pick up pieces and still fly. And then you scrape the leftover crumbs into this box for the fish. . . . Hey, are you getting this down in your little notebook?"

PREDICTABILITY AND ROUTINE are the primary currencies at the chapel. First one here every morning, just as the newspapers thump on the still-darkened doorstep, is sixty-four-year-old Murray Prescott who lives with his wife, Barbara, a few doors down. Murray is a tall, kind man with a soft, polite demeanor that belies the occasional "damn" or "honey" that sneaks out when he gets worked up talking to me. Every holiday he decorates his home like a glorious funhouse with appropriately themed lights, signs and outlandish blow-up dolls; recently a fifteen-foot inflatable turkey from Wal-Mart was added to bolster the Thanksgiving collection. Barbara, meantime, is a spitfire who insists on listening to Christmas music in July, likes to order Murray around and eats a grilled cheese sandwich on white bread whenever she pleases, diabetes be damned. Murray and Barbara met at a roller-skating rink in 1955 and were married soon after. Nine years later, they moved to Garrison's Landing where the Guinans quickly became their best friends, with the children running between houses for dinner depending on what was being served. Both the Prescotts have filled in behind Guinan's counter through the years, and Murray still shows up on Sundays to help John piece together the fat *New York Times* and bemoan changes in the town.

At some point after Murray comes Lou-Lou, who is actually part black-and-tan coonhound and part beagle. She eats her way down the river landing, slipping into the river for a swim if it's hot and making Guinan's her last stop. Most mornings she shows up between seven and eight, squeezes her wide frame through the screen door behind some commuter and butts John in the legs. He hands her a roll, pats her on the head above those bulging brown eyes, then with a gentle nudge tells her, "Lou-Lou, go home."

And then there's old Ken Anderson, who lumbers in like clockwork around 7:20 A.M. "GOOD morning, John," the voice booms across the store. Ken is six feet three inches tall, and drives his little red Ford Fiesta straight down the middle of the landing. Even chunky Lou-Lou moves quickly when she sees him coming. When Ken parks, he leaves his engine running, trying at age eighty-eight to save time where he can, I suppose. But I learn never to underestimate the sharpness of his mind. Once I ask him how long he's had his car, and he looks at me, disappointed, and retorts: "Since I bought it."

"Morning, Ken. How are you?" John typically goes. He pulls Ken's paper from the Reserved stack, which he has lined up neatly on the ice cream cooler, last names scribbled in black pencil: Anderson, Banker, Chip, Connie, Cuneo, Cutler. . . .

"Moderately well, thank you," Ken replies. This is one of two standard answers, the other being, "Just like two pianos, upright and grand." White kneesocks glare from beneath his khaki pants, which are inexplicably rolled up higher than necessary. "You know," he says, leaning against the metal counter, "I didn't get my *Putnam County News and Recorder* yesterday. Some guy was taking too long looking through the newspapers. I asked him: 'What do you think this is, a library?'"

Ken ambles toward the door, where he slides the store's OPEN/CLOSED sign to OPEN, then folds his frame into the tiny red car, guns the engine as a warning and backs up.

THE COMMUTERS PASS through, a fast-moving parade of well-tailored suits, bags and anxious demeanors. They dodge one another, reaching for the coffee, tea, newspapers and making their change. After a while, they are all a big blur to me, but John gives them flesh and divides them into two camps: those who are "good guys" and those who aren't.

". . . And that's Ray, he's some kind of big shot at Morgan Stanley. Good guy, though. Gave me a DVD about Shackleton's explorations. His wife, Patty, is a sweetheart, works in the city *and* raises their kids. His brother, Jimmy, lives around here too—really good guy. Now the guy with the fur cap is shifty, stay away from him. Same with the woman driving the fancy sports car; she always parks illegally in front of the store. Frank Geer there, he's the Episcopal preacher from up the road at St. Philip's Church. He drops off his wife, Sarah, for the train and then reads the newspapers with his coffee back in the bar. Only pays for the *New York Times,* but that's all right because he's the preacher and a good guy and puts the others back when he's done. And that bearded fellow over there, Finnegan, we do landscaping up at his house. Nice guy—comes to Irish Night. Patricia there, she's the best dresser.

"Okay, see that smaller wavy-haired guy? That's Chris Davis. He owns the golf course. Runs some big mutual fund company in the city."

That's Chris Davis? I say. Of Davis Funds? You're kidding me. The guy manages billions of dollars. He's a huge deal on Wall Street.

"Yeah, well, I guess. But he's a good guy anyway. So there's Ryan Cuneo—that's one of Dad's old golfing buddies, you know, the Fearsome Foursome. And Don Abel, he works up at the town recreation department. . . ."

DOWN AT THE store, Jim's drinking his morning tea when I walk in with the bonus puppy for the first time. She's tugging at her leash and sniffing around for crumbs. Eventually the puppy's nose leads her into the kitchen, where she puts her paws up onto Jim's chair and peers onto his plate.

"Well, well," he says cheerily. "What have we here?"

This is Dolly, I tell him, testing out the name we picked out the night before, Dolly Levi Garrison. Like Barbra Streisand's character in the movie, I add, with Garrison tacked on.

"Well, hello, Dolly." He laughs. "That is a fine name." Dolly gives a big yawn in his face.

"They yawn a lot at this age, you know," he says, rubbing her cheek and giving her a nibble of his toast.

Margaret walks in carrying laundry and groceries and Dolly jerks away from me and runs over to her, sniffing her boots. I hurry to get the puppy out of the way, but Margaret sets down the bags and bends over to stroke Dolly, who leaps up and licks her ear. The detective points to Dolly's feet. "Big paws," she says. "You're in trouble."

I show her a couple of pictures of Dolly running around the yard. She takes time to look at each of them, and then picks one from the pile and sticks it on her dad's refrigerator.

BAD WEATHER CHANGES everything at the store. When the rain spreads in sheets across the hillsides and river, the temperament of the commuters spirals downward. The gloom makes everyone sleepier, so people linger in bed just a little longer. The extra few minutes mean they are already running late by the time they pull the car out of the garage. If it's coming down hard, the dirt roads are an obstacle course of muddy potholes and washed-out corners. Add slower traffic and the stress of staring through the beating windshield wipers for deer, and by the time everyone rolls into Guinan's, they're already stressed out and the workday hasn't even begun.

I watch them come in, stamping wet feet on the soft wooden floor, shaking damp heads and juggling their gear as they squeeze by one another, tense and trying to get a cup of coffee before the train's up. John——who's had less sleep than most of them——cracks jokes, compliments colorful raincoats and doles out extra umbrellas to those who forgot theirs.

The clock in the bar is set ten minutes fast so folks sitting with their coffee won't cut it too close and miss the train. But on days like these, even that's not going to help the chronic stragglers. Without fail, there's always one or two who rush in at the last minute, whacking everyone else with their bags and breathing through their nostrils like charged-up racehorses. They'll start to pour their coffee just as the train whistles outside, at which point they turn to John, panicked, fumbling for a wallet, sloshing coffee on the floor, newspapers slipping from their arms.

"Go," he says to them. "Just go and pay me later." They nod, eyes wide, and rush through the door and out of sight onto the platform.

"It's always the same people," he says, shaking his head, mopping up the mess they've left behind. "All that scrambling over each other, slipping on their hard shoes, rushing for the train, you gotta ask yourself, Was it worth it? They couldn't leave five minutes earlier?"

I don't answer. Because if I did, I'd have to tell him that in those frantic latecomers, I see myself.

AFTER A FEW weeks, John and I settle into a routine of sorts. If I arrive in time, I'll wrap the fresh doughnuts and bagels that get delivered each morning, slice five rolls (don't ask why five, that's just the number they've always done), spread them with butter and try to keep the coffee going. Meantime, John writes up the newspapers, takes out the trash and mops the floor in back. We both handle customers, though I constantly have to check price lists (Yoo-hoo, 85 cents, bagel with cream cheese, 95 cents) or ask John for help tallying long orders.

One morning I start poking through the items scattered on the counters and in cabinets and find, among other things, a plastic yellow Easter egg, bits of spare change in glass jars, a single unattached sunglass lens, bottle caps, something labeled "holy water," rubber bands, 39-cent bobby pins, a sparkly barrette, a pink princess doll, as well as a dusty $1.53 bottle of pink Alberto VO5 herbal scent conditioner.

And that's just what's in front.

From further in the back, I unearth Drano and something called Vaseline Hair Tonic and Scalp Conditioner—1.75 fluid ounces, manufactured by Chesebrough Ponds Inc. *Directions: Pour just a few drops (no more than four) into hand. Massage vigorously into scalp. Then groom.* Cost: 99 cents.

When John passes by carrying a plate of butter, I hold up the hair tonic with a questioning look.

"Hey, you never know what you'll find in the Archives," he tells me.

Then I unearth a couple of dog-eared ledgers held together by duct tape and rubber bands. When I show them to John, he blanches. "I'd be careful with those."

What are they?

"Dad's account book for customers. Goes back to when he first opened the store."

I creak one volume open carefully, and sure enough, the first page is dated 1959 with customer names running in columns, the newspapers they subscribed to and lists of charges added and crossed out. You can total how much the Belcher family spent in 1963 or, in volume two, find a note stating that as of Aug. 3, 2000, Mary Ellen Yannitelli wanted her sons' charges limited to $1 per visit. There are notations if someone was sick or on vacation and for how long.

This is incredible, I tell him. I mean, the whole town is in here practically.

"Yeah, well, Dad never throws anything out," John grouses. "We should open a goddamn museum." He says it gruffly, but when I go to put the ledger back, he calls over his shoulder: "Hey, now, be careful with that, okay."

I GET FASTER. Can wrap rolls and cut all the old bread up before John finishes with the newspapers. Multitasking now means taking orders while simultaneously buttering a bagel, calculating change and remembering someone's name as they walk through the door.

Frank the preacher tells me tidbits of town lore—some more factual than others. For instance, he says, the castle on the hill named Castle Rock was built by a wealthy robber baron in the 1800s—that much is fact. However, some say he chose that hill because the earth's magnetic pull created such a powerful energy vortex there, and he wanted his legacy erected where the eyes of passersby would inevitably be drawn. Maybe that part is legend, but from then on, every time I pass the castle on the road, I feel myself glancing up toward its tower, wondering whether it's my own free will or some underground force at work.

Frank also gives me a book he cowrote: *Where Was God on September 11th?* During lulls in the crowd, I'll sit back at the bar sipping coffee and reading passages while the sun spreads across the river.

". . . the disaster itself had a ripple effect. The pain and the horror and the tragedy of it spread out through the world like the ripples on a pond when a pebble's been thrown in. But the healing travels back the other way, like the same ripples coming back towards the center after they've reached the shore."

One morning two swans strut by the back staircase and I holler at John to come look.

"Eh, that's probably just some of Fionnuala's offspring," he says, nonchalantly.

What do you mean, Fionnuala? I demand. Who is Fionnuala?

"A swan," he says, as if it's perfectly natural to have a swan hanging around. "What, you don't believe me?" He strides into the kitchen and pulls a photograph off the refrigerator. Sure enough, there's a giant white bird standing at the store's front door with her head peering in the glass.

"See," he says.

Later I find out that Fionnuala is named after the daughter of Lir, god of the sea in Celtic mythology. According to legend, she was transformed into a swan by her stepmother and condemned to wander along the rivers of Ireland until Christianity came to the island.

Mornings used to be about hitting the snooze button one last time and racing across the street into the office with my hair still damp. Now sometimes I'm up with the sun and cavorting with a coonhound and a mythological swan's progeny before work.

This is great, I think. I could do this every day.

OKAY, I TAKE it back. This sucks. How do Margaret and John *do* this? I'm only here a couple of mornings a week but even so, the schedule is getting to me. I yawn wrapping rolls in the morning and slug back cups of coffee until my stomach burns. The *Wall Street Journal* has now moved most of us from New Jersey into the temporary New York City offices. On the days I commute to Manhattan, I arrive at work around 10:30 A.M. at which point I already feel like eating lunch if I've spent the morning at Guinan's. By 2 P.M., I find my head dropping down at my computer monitor and want nothing more than to crawl back into bed.

I calculate the amount of money the store takes in between 5 A.M. and 6 A.M. and estimate it to be only around twenty to twenty-five dollars. This seems highly inefficient—obscenely long hours are fine if there's a payoff in sight, like a raise or promotion, but the Guinan kids are not getting paid at all to be here. So I decide to broach the matter with John.

I don't get it, I say one morning while we chop up bread for the birds. It's not like you make a lot of money in that first hour. Why not just open up at five-thirty or six and get some extra sleep? You'd be more productive the rest of the day.

He fastens his blue eyes on me, and for a moment, he looks remarkably like his father. Then it passes, and he pats my shoulder. "I don't know. I guess I don't want to be the one to tell those guys on the 5:09 train that they're gonna be standing out in the dark without a cup of coffee or newspaper."

I open my mouth to respond—to tell him to be practical, to think like a businessman. But he's already moved on and is studying the day's inventory lists. So I go back to chopping.

THE 509ERS, AS they're called, are the Navy SEALs of commuters. They are either the bravest or the craziest members of the working world. Their days begin in the predawn hours when country dark is heaviest. Out here, most roads don't have streetlights. When late-night traffic has ceased, there are no perpetually vigilant skyscrapers or careening taxicabs to lend the aura of company.

None of this fazes Stacey Gibson, a manager with a major New York pharmaceutical company and the equivalent of the 509 crew's command master chief. At 2:30 A.M., about the time many New York City bars have just closed up, fifty-one-year-old Gibson's alarm goes off on Snake Hill Road. The beep is loud and jarring to ensure she can't sleep through it. The first thing Gibson does is to walk her five dogs: Patti, Evie, Max, Sam and Luke. Luke, who is slightly more high maintenance, gets fed first and then locked into the mudroom until he settles down. The other four dogs lumber into the living room to chew on bones. The second thing Gibson does, from 2:45 A.M. until 3:45 A.M., is to walk on the treadmill at a brisk rate of four and a half miles per hour while flipping between CBS and ABC news. Then she takes a shower, dresses and at 4:50 A.M. reports to the little chapel on the river.

There she is joined by the rest of the troops, among them a hunky New York City cop and a doorman for the Palace Hotel. If the papers have arrived, the 509ers help John carry the stacks inside. The hour is odd, disconcerting. Everyone there knows it's not exactly normal to be awake so long before the sun comes up. Here in the store, though, they are not alone. The lights are on, a pot of fresh coffee is brewing, and they can talk in regular voices. Here, the early hour is just a little less strange.

Around 5:08, they move to the train platform, splitting up now, individual bodies illuminated by the old-fashioned green lampposts and their wan yellow light. At 5:09 A.M. sharp, the faint hum of an engine sounds. Then comes the high-pitched stinging along the tracks as if thousands of tiny pushpins are being sprinkled upon the rails. The train's horn sounds once coming through the rock tunnel. And then the metal beast is upon them, white headlights shining like four eyes, earlier prey in her belly. The doors sweep open; she waits for the next round to file in, then swallows and is off, orange lights blinking from the rear.

There are unwritten rules at this hour; nobody talks; most people sleep. Stacey Gibson usually works or watches dawn spin cotton candy pink across the Hudson River. Moving south, the silver mass of metal muscles her way past the reeds and cattails, through surburbia and then hits the urban asphalt and graffiti-riddled tunnels before finally stopping at Grand Central Terminal. It is there that the 509ers deposit Guinan's Styrofoam cups and buttered-roll wrappers into the blue waste bins and emerge into the frenetic streets.

The time is 6:15 A.M., almost four hours since the alarm went off on Snake Hill Road.

I BEGIN TO practice basic arithmetic in my head. I do it while driving, borrowing random numbers off billboards and license plates. I add on the subway and in bed before I go to sleep. Since I'm clearly not a natural at this, I devise ways to get the total most easily. Anything in a 25-cent increment is good. With numbers that aren't, I round up and then subtract: so, for instance, $.85 and $1.25 will become $2.25 minus $.15.

$2.10.

Okay, I know. This is basic grade-school stuff, right? Except that it's not when there's a line of late commuters staring you down. Or when some hairy guy is yelling across the bar because you gave him a Ballantine and not a Rolling Rock—as happened the first night I ended up filling in for Jane.

Sure, as long as my computer or Palm Pilot calculator is handy, it'll be fine. But that's just it. Watching the Guinans whip through numbers, I'm jealous that they aren't dependent on some machine. When children come through the store to buy candy, Margaret and John help them calculate their own change before giving them a free pretzel for the effort. I begin to wonder if all my smart gadgets have actually made me stupid. How can I complain about the speed of my broadband connection when I can't even calculate the total of a quart of milk, bagel and orange juice in less than five seconds? Plus, there's something comforting about the fact that if the power went off, and there were no spare batteries or generator around, then Guinan's could just keep doing its business. And so I keep practicing.

One day I'm at a deli in Manhattan buying lunch and plunk my sandwich, chips, banana and Coke onto the counter. The cashier begins to ring them up, same time as I'm silently adding out of habit: $5.75, $.50, $.80 and $1.25 . . . becomes $7.50 plus $1.00 minus $.20 is . . .

"EIGHT DOLLARS AND THIRTY CENTS," I holler out triumphantly before he can announce my total.

The other customers in line look away uncomfortably while the cashier bags my lunch with a haste that suggests he's worried additional outbursts might follow.

I'm just practicing my addition, I tell him.

"Uh, sure," he says, glancing over my shoulder. "Next in line, please?"

IT'S THE LITTLE things that can make an otherwise smooth morning turn sour.

"Excuse me, miss."

"EXCUSE me, miss."

I look up from the counter where I'm slicing a bagel. The place is swamped with customers banging into each other in a rush to make the 8:30 train. Before me stands an older woman, no one I've seen before. Her skin is pale, veins shiny at the surface where it's soft, no doubt from decades of expensive face cream. She is prim, back arrow-straight, and she's clutching a cup of coffee with one pinky out to the side as if she were at some tea party.

"I'm just letting you know that it might take more than a paper towel to clean up over there." She nods with a frown toward the coffee station.

Clean up? I ask, glancing over that way. I had just wiped everything down about thirty seconds ago.

"Well," she says, "a sugar packet malfunctioned when I opened it, and there's sugar all down inside the coffeepots and burners."

Now I can smell it, a sort of gaggy sweet burning scent. I stare at her for a moment.

Malfunctioned? I say finally.

"Well, it sort of spilled everywhere when I opened it." She looks annoyed that I haven't apologized or rushed to the scene. "I just thought you'd like to know so you could clean it up."

I look up at the line of people trying to squeeze the last drops out of the one unfouled pot. The train's up in about forty-five seconds. Now half of them won't get coffee. And what I want to say is, MALFUNCTIONED? WHAT DO YOU MEAN IT MAL-FUNCTIONED? YOU MEAN YOU WERE IN A HURRY AND SPILLED SUGAR EVERYWHERE AND NOW YOU'RE ACT-ING LIKE IT'S MY FAULT. I'M SWAMPED OVER HERE. WHO DO YOU THINK YOU ARE?

She's a customer. That's who she is. End of story.

Thanks for letting me know, I say politely. And I pick up a wet rag and head toward the burning scent.

DOWN THE BACK stairs next to the river sits an old rusty red metal barrel with a few holes poked into its sides. Rather than clog the garbage with the pastry boxes and other trash that comes into the store each day, the Guinans take it all out to the barrel and burn it into a fine ash.

"I put this lightweight brown paper in there first because it lights fastest," John tells me one morning. "Be careful, don't get downwind from the flames." He watches as I poke at it with a long charred stick until he's sure I'm not going to set the house, or myself, on fire. Then he leaves me alone with the boxes and the river.

I begin lifting each box with the stick, setting it down lightly into the barrel, knocking it a little to make it settle in. The heat, as it rises, makes hazy the space between the barrel and the foggy river, as if I were staring through distorted glass. Everywhere, there is life moving forward. To my right, the seagulls bob, hoping for a second breakfast. A long barge pushes upriver, slow but de-termined, while a cargo train rumbles by underneath West Point, its colorful cars tiny and toylike at a distance. The tree leaves shine bright lemon-lime, a newborn green that will darken in the weeks to come.

From now on, this is what morning will mean to me. A burning barrel coming to life while the fog peels off the river like a coat.

⌇

I STARED AT the painting, trying to make out if the signature really said Picasso.

"You like this piece?" the designer asked me. I looked over at Donatella Versace and nodded. It was December of 1997, and the fashion icon was hosting a group of financial reporters for dinner at her Manhattan apartment. There were bodyguards outside the front door, a not-so-subtle reminder that her brother Gianni had been gunned down only five months ago. It was a good thing anyway about the guards; I'd never seen so much beautiful art in a place that wasn't a museum.

The Versaces were considering taking their empire public on the stock exchange, and as the Wall Street Journal's *designated fashion reporter, I was seated next to Donatella.*

"We met once, right?" she said warmly, her trademark platinum hair glowing against her bronzed skin. "At a fashion show?"

Well, it was very brief, I said, figuring she must be mistaking me for someone else.

"Oh, yes, we met," she replied smoothly as the waiter spooned caviar onto my pasta. "I remember thinking to myself, now that is a Versace girl."

"A Versace girl" . . . It sounded so heady that I wanted it to be true. So I sat up straighter, twisting my pasta around in a spoon, and tried to ignore the little voice inside me snickering and whispering that I was really more a Levi's girl and trying to forget it.

the cadet

It was one beautiful, warm time.

Occasionally on weekend afternoons, I'll sit back on the worn rose-colored couch with Jim while he watches TV. Like his son John, the barman says what comes to mind, mostly stories from escapades past. The river is full of traffic now, which reminds him of the days when he took folks by boat over to the West Point Army football game. Thirty dollars bought you a ticket, food and beer for tailgating, which Jim hauled over from his ready stash down in the cellar. The stadium itself was dry, but that was no matter as the women slipped liquor bottles into their bras to keep the party going throughout the game. When Jim and his entourage entered—at one point there were nearly two hundred of them—the scoreboard lit up announcing that "Guinan's from Garrison" had arrived.

"It was a good time." He smiles, remembering. A moment later he adds offhandedly: "Where there's life, there's hope."

THE SNIPPET OF river between Jim Guinan's window and West Point had long been a critical passage. As America had moved toward Civil War, Garrison's Landing became the main connecting point between the train and ferry that carried travelers to the West Point military base across the river.

One cold December evening in 1853, a knock at the door awoke the ferry's captain, Jesse Austin. There before him at the cottage door stood Robert E. Lee, then superintendent of West Point, demanding to be rowed across the river to the military quarters. The nervous ferry captain tried to dissuade the determined Lee; it was a stormy night and the river was filled with ice floes. But Lee would hear none of it, and so Captain Jesse reluctantly readied his vessel. For an hour and a half he navigated the treacherous Hudson waters before reaching West Point. At this point, it is said, Lee took the exhausted ferryman to his quarters, where they shared a good-sized bottle of spirits. Yet another night, in June of 1862, the ferry captain received a telegram from the U.S. War Department telling him to keep up steam on the ferry throughout the night and await further orders. Somewhere between 3 A.M. and 4 A.M., a special train stopped in Garrison and President Abraham Lincoln stepped out. He was met by Henry White Belcher and a few other prominent townsmen, and then ferried across the river to a hotel just south of West Point.

But it wasn't just national security or politics that inspired men to make the treacherous late-night journey across the Hudson.

BACK IN THE 1960s and 1970s, when military life at West Point was stricter, Guinan's was something of a halfway house for cadets. Those embarking by train for a few days of R&R would take the ferry in the warm months over to Jim's docks. Since they often couldn't leave the post without wearing the West Point uniform, the Irishman invited the young officers-to-be into his living room and let them change into their civilian clothes there—a tradition that continues even today. Most cadets simply left their military attire spread in neat piles across his green-carpeted dining room floor, waiting to be reclaimed at the end of a weekend. Sometimes Jim lent them money or Peg gave them free sandwiches for the train. On parents' weekend, the cadets marched their mothers and fathers into Guinan's to introduce them to this couple that had become surrogate parents.

More adventurous, however, were the young men who risked paying an illegal visit to the bar at night. Even though the legal drinking age was still eighteen in the sixties and seventies, cadets nonetheless were prohibited from drinking within twenty kilometers of the post. For the brave few who risked getting caught, the most preferred means of transport to Guinan's was by rowboat, one commandeered stealthily from the West Point docks at night. Perfected, and tides accommodating, the journey took twelve minutes, Jim says. "Except for the time someone stole the lads' paddles," he explains. "Then the four of them used sticks to propel themselves and it look a bit longer."

If someone at the bar spied the cadets coming, Jim would wait down by the docks, help pull the boat in and then usher the rowers inside. And for a few hours, these young men tasted freedom in the flavor of beer, darts and unsupervised male banter. The barman loved them all and welcomed each equally. But, as he tells me, there was one young man in particular. . . .

⌐

MARCH 1977.

A young West Point cadet crouches in the bushes with a fellow classmate, staring intently at the neon signs on the opposite bank. Night is falling and the red-haired boy, a husky plebe named Tom Endres, is bored and restless, weary already of the regimen of military life. He's only come to play lacrosse and to live down the words of his father, who told Endres when he graduated from high school with his long hair and well-honed disdain for authority: "West Point? You won't be accepted there. It's a man's school."

Right now Fourth Class Endres is looking for a diversion and assumes, correctly, that the neon signs across the river belong to a bar of some sort. Endres is nineteen, old enough to drink legally, but that's of no importance by West Point rules that prohibit drinking near the academy when off-post. However, the headstrong cadet figures that by putting a river between him and his instructors, he's probably in pretty safe territory. So as the sun dips below the hills around West Point, Endres coaxes a buddy down to the academy's yacht club, where they commandeer a dinghy and slip the boat through the weeds and marsh until they are directly across from the twinkling neon lights. Then the pair begins cutting across the river in silence, save the rhythmic dip of the oars rippling through the water.

Their plan might have gone smoothly, except that Endres underestimated the currents of the Hudson. He's rowing as hard as he can, but the neon signs are slipping past him. Mischievous as always, the river carries the cadets past their mark and finally dumps them ashore at Arden Point, several hundred yards downriver. Sweaty and disheveled, Endres ties the boat to a tree stump and the exhausted boys make their way through the muck and up along the train tracks until they reach Jim Guinan's door.

Inside is a crowd of about eight to twelve men. Some are playing darts. The cold, dirty cadets traipse into the back and size up their surroundings. The crowd pauses for a moment to examine the new arrivals, red-faced and breathing hard in their muddied army-issue sweats. It is the bartender who speaks first.

"Well, well," says Jim Guinan with a grin. "I'm glad to see there are still some cadets who have the balls to sneak over here."

CADET ENDRES HAS a beer. Then he has one more. And one more, and well, just one more until his 1:00 A.M. curfew is fast approaching and he's in no shape to row the boat back across the Hudson with those quick-moving currents. Jim Guinan realizes this and corrals a well-built and relatively sober guy named Don to drive the boys back over. He pats young Endres on the shoulder and hands him his phone number. Tells him rowing over is fine and shows the proper spirit, but maybe he should just call next time he needs a lift.

The next day, Endres feels great, although he's a little concerned because he left the rowboat tied to a stump just south of Guinan's. A day and a half later, he tentatively makes his way back to the West Point boathouse to see if he can spot the craft across the river. And there, to his surprise, lashed up snugly in its rightful spot, bobs the stolen vessel. Somehow, the Irish bartender has managed to have it returned with no one the wiser.

Cadet Endres stops rowing across the river. But he doesn't stop going back to see Jim Guinan. Sometimes Endres sneaks out with upperclassmen in their cars. For a period during his sophomore year, he hides his own car illegally around the post. In a real pinch, he telephones the bar and Jim sends someone to pick up him and a few pals at the academy's gate. Jim always remembers each of their names and treats them as equals. But that doesn't stop the boys from getting cocky sometimes. They get going playing darts and empty a few beers. And then they start boasting about how they can beat the Irishman at his own game.

One night Jim lets them get good and riled up, and then when he's got the attention of the crowd, he stands up and walks around the end of the bar, swaying a little since he's had a few himself.

Endres and his buddies are grinning, sure they've got him beat this time. And then, POW. Jim lands a dart right in the bull's-eye. He leans over to the silenced boys and tells them it's important to be able to back up your boasts. That you don't want to be someone who "has flies on 'em."

That is to say, don't be a bullshitter.

Endres never gets caught going to Guinan's. And he actually stops hating military life quite so much. Or feeling like he needs to prove anything to his father. He leaves West Point's gates, becomes a pilot and flies with the 101st Airborne Division. But even after he makes his way through Korea and Somalia—even after he sees the world—he never forgets Jim Guinan or Garrison.

Cadet Tom Endres goes on to become Colonel Thomas Endres. And in 1990, he comes back to the Hudson Highlands with his nephew, son and wife, Sandy, to visit some relatives. The colonel decides, for old times' sake, to stop by Guinan's. Warns his family as they are approaching the screen door, "Now I don't know if he'll remember me. It was a long time ago."

The door swings open. Everything looks the same. Wood floors still trampled so tender they almost feel like dirt. Compressors running. Candy counter overflowing. And there, in the back, the colonel sees Jim sitting on the far stool leaning against the wall, his head hanging slightly, body backlit by the sunlight so that he's just a silhouette.

Endres steps into the doorway, and the noise makes Jim look up. There isn't even a pause.

"Ahhh," says the barman. "I knew you'd come back, lad."

WEST POINT GRADUATES only owe a minimum of five years' service after graduation. Yet it's now been almost a quarter of a century since the cadet who never wanted to attend West Point went on active duty. Even more ironically, perhaps, Colonel Thomas Endres now works back at West Point as director of all cadet activities. He's grown into a charismatic, barrel-chested man who's built neat and muscular like a spark plug. His blondish-red hair is cropped into a sharp military flattop. The colonel takes his job seriously—manages a thirty-million-dollar operation, running 110 clubs and teams, six restaurants, all entertainment and everything extracurricular the cadets do. But on Monday afternoons he'll hop in his silver Porsche and slip across the river to the little green bar of his youth where everyone just calls him Colonel Tom.

If it's summertime, Colonel Tom will have his own boat tied up (legally this time) down at the dock under Jim Guinan's window. The colonel's own home has become something of a halfway house for cadets these days, and many afternoons he and his wife linger around the bar waiting to bring a few bright-eyed boys to their place for a home-cooked meal. Sometimes, if he's feeling up to it, Jim will come out to sit at the colonel's elbow. In front of the two men is a picture tucked into the corner of the bar's cracked mirror. The image captures a baby-faced cadet, Endres, seated at the bar, impish grin spreading over an empty glass, as a much younger Jim Guinan stands nearby raising his full glass to the camera.

Colonel Tom's not one for bullshit. No, he learned that lesson well. But ask him what he thinks about his days of endless beer and darts and tales and he'll tell you this and he'll look you in the eye.

"It was one beautiful, warm time."

LIKE MY GRANDFATHER, my godmother, Gwendolyn, had her habits: drank martinis every night, cursed when appropriate and smoked as much as she goddamn well pleased. A book editor at the University of North Carolina Press, she grew her blond hair down to her thighs and corrected people's grammar without apologizing. I liked her very much when I was young; I suspect I'd have liked her even more now that I'm older.

When my godmother got sick, she withdrew from us all. Didn't want to sit around and pretend to have a Merry Christmas when she wouldn't live to see the next one. Back in New York City, I did what I thought she'd like. Wrote a letter just talking about average things in my twenty-seven-year-old life and telling her I loved her. I checked the spelling closely before sealing it shut.

I meant to mail it that afternoon, but something happened——I got busy, I guess——so I sent it the next morning. What was one day? I figured. Except that, of course, it was everything in this case and the letter arrived the morning after she died.

Had she been able to, I suspect she'd have called a spade a spade—— which is to say that my excuses, whatever they might have been, were bullshit.

the ride

Pace yourself, kid . . . it's not a race.

God, I swore I wouldn't do this to myself again."

John is grimacing, leaning over and rubbing his legs. It's seven forty-five in the morning and his eyes are bloodshot from lack of sleep. We're in a lull between trains; I'm on Duck Detail, slicing up the old doughnuts.

Wouldn't do what? I ask him, carefully scraping the crumbs into the carton.

"Not be in shape for the ride," John says, twisting his back slowly. "Haven't been on a bike once, and we leave in a couple of weeks."

He's going on the Northeast AIDS ride, a grueling 323-mile bicycle trek from New York to Boston to raise money for treatment of the disease. It's his seventh year riding, a strange commitment for a man who once wouldn't even sit next to a gay man at dinner—"I don't know, it just made me nervous. I can't explain it any better than that," he says.

But his attitude changed with a single phone call that came one night in 1994. John and his wife, Mary Jane, were watching TV in their bedroom when Mary Jane's mother rang. As his wife held the phone listening, John watched her face go ashen.

She hung up the phone and turned to him: her baby brother, Tommy, had AIDS.

He'd been sick since 1985, flying back and forth to Montana to help care for his partner, who was dying at home with his own parents. Mary Jane had picked up on little signs: Tommy's thin frame, his long trips away, the absence of a girlfriend for so many years. Somewhere in the back of her head, the suspicion lurked—AIDS was all over the media—but the notion was so horrible she pushed it away, willing herself to be wrong.

But now, there it was, awful and alive in her bedroom. After Mary Jane hung up the phone, John sat on the floor holding her. As she sobbed into his shoulder, he began thinking of all his macho behavior toward Tommy and all the things he should have said and done but never did. And he felt sick to his stomach.

"What are we going to do?" Mary Jane asked him rhetorically, still in shock.

John swallowed. Strangely, the answer seemed simple enough right then. "We either walk away or we have to stand behind him and support him however possible. And that's what we'll do."

For John, however possible came on the day Tommy asked him to do the AIDS bike ride with him. They rode together until Tommy couldn't; then John rode alone on Tommy's bike. Mary Jane volunteered on crew most years, helping with gear and food distribution for the thousands of riders. Tommy died in 2000, and while Mary Jane and John had no way to reclaim lost time with him, they did what they could with the time they had left and were by his side until the end. John still does the ride each year, pushing through the three hundred plus miles even if he hasn't sat on a bike since the year before.

So why ride if you're not ready? I ask. Why not just give them all the money you've raised and stay here?

"Why ride?" He gives me a disappointed look. "Look, kid, I made a promise. You do what you've got to do. And I'll do this until I can't anymore."

A FEW DAYS before John is scheduled to leave, I take my mountain bike out and pedal slowly through the hills around Garrison, dropping down into Cold Spring and then heading back home. In the cool Highlands air it feels pretty good. After clocking about 13 miles, I make a decision. What's the big deal, I think. If I can ride 13, I can ride 323.

I'm going with you, I announce to John the next day. For moral support, I tell him.

He breaks into a grin and gives me a hug. "Really? Are you sure?" he asks. "It's a long way."

Oh, it'll be fine, I assure him.

DAY ONE. MILE 18. Another 305 miles to go.

My legs are on fire. My knees hurt. I don't know what I was thinking. It was one thing gallivanting around Garrison, but now I've got this hydration pack on my back, and people are passing me in a constant stream. People, I note unhappily, who look to be in a lot worse shape than I am. They smile and give a little cheery shout——"On your left"——as they glide by. I glare at their backs and shift my aching buttocks.

"Pace yourself, kid," John says to me at the second pit stop when he notices my legs shaking. "It's a ride, not a race. It doesn't matter if you finish. They'll pick you up if you don't make it. Just have a good time."

But I don't know how——not to finish, that is. That old internal clock has kicked in, and somehow I think that not finishing is synonymous with failure. Projects, stories, deadlines——completion is the only measure I know of a job well done. More than that, though, I don't know how to just relax and enjoy the process. No one is in competition here. And yet I'm still competing with that stupid clock. So I push through every day, allowing myself only a few minutes at rest stops, pedaling until I have blisters between my legs and nearly collapse off my bike at the end of our 100-mile Day 2 journey. I start early each morning, and if John's not ready, I don't wait for him, afraid that even a fifteen-minute delay will foil my ability to complete that day's ride.

John, on the other hand, takes it easy. He chats with people along the way; stretches out at the pit stops for lunch and rests for a while in the sun with other riders. When a stranger gets a flat tire, it's John who's off his bike helping patch or put a new one on. One day he stops for a piece of pizza and beer in an old roadside shop just because it reminds him of home. When he arrives at camp tired each afternoon, he still makes time to help set up the other riders' tents. He does all this, and still completes every day but the 100-miler—and that's only because he stayed behind encouraging his tentmate, an overweight fellow whose knee was hurting.

The last evening, after we get our food, the organizers ask us to take a moment and pay tribute to riders who are no longer with us. John bows his head in silence. After he looks up, he wipes back the tears without a bit of embarrassment. He signs up for next year's ride right then. People keep coming up to him, recognizing him from past rides. They pat his shoulder and affectionately call him "breeder" because he's heterosexual. He laughs and puts his arm around everyone, all the old macho nervousness long gone.

Then it starts to rain, and even though we're both beat and ready to hit the showers, he grabs my arm.

"Come on, kid," he says, pulling me up from the bench.

What? I ask him wearily.

"There are still riders coming in," he says. "Let's go cheer them on." And so we lug ourselves up to the muddy hill, and stand there as darkness falls, clapping while the final few riders push through, many crying with exhaustion and joy. A couple of people who are paralyzed or missing one or both of their legs come in on cycles they are pedaling with their arms.

"Hey, iron legs," John calls to me.

I look over at him.

"Thanks for coming with me on this trip. It means a lot that you're here helping Mary Jane and me honor Tommy. And you've done great, kid."

He hugs me, and I look over his shoulder up the hill at the rows of little blue tents filled with bodies that ache from pushing for someone besides themselves. It is then I realize that while I may have finished every mile, John is by far the better rider, and person, of us both.

WHEN WE GET back to Garrison, John tells Margaret he needs to take a break from the store for a while—needs to think through some things and spend more time with his wife. His cousin Kathleen is in town to help for a week or two so he doesn't think it'll be a problem. But his father starts getting up at 4:30 A.M. again for the early shift, standing on his feet until Margaret and Kathleen show up to relieve him. One morning I walk in with Dolly around 8 A.M. and find Jim sitting in the kitchen, cheeks unshaven, smiling eyes a little glazed. I haven't been down to the store in a few days and am surprised to see how frail he looks.

"Me foot, it's bothering me a lot," he says before I can ask.

Is it swollen? I ask.

"Ya, swollen, the toes are curled up like this," he says, making a ball with his fist. "I got an appointment with the doctor tomorrow morning. She's going to give me hell. It's not so much the getting up early. It's the standing all day. That's what gets to me.

"Since John went on his bike ride, that's when I took back over," he continues, sipping his tea. Jim crumbles a bit of his buttered roll into his hand and feeds it to Dolly, flat-palmed.

Do you want John to take over here at the store for you? I ask Jim. Then I add: Permanently, I mean.

His face is unreadable. "John has to make up his own mind," Jim tells me. "He can do anything, you know. The store, or anything he wants. He's got a lot of talents."

The Irishman's words stop me. Whenever John and his father are in the same room, they are usually carping at each other about this and that. Jim tells John he talks too much. John tells his father he doesn't listen enough. This is the first time I've ever heard Jim praise his eldest son out loud.

Have you ever told John that? I ask him. That you think he can do anything?

The Irishman waves his hand brusquely. "Don't need to tell him, he knows it."

Just then, a woman comes into the store with her young son and stacks a bunch of goods up on the counter. Jim gets up slowly and runs through her purchases, adding out loud as he goes: "$1.25, $5.95. . . ."

"Wait, you're charging me too much," the woman says, interrupting him.

So he starts over, but she does it again——"Wait, how much are you charging for that?"

"Ma'am," Jim says, "please don't interrupt while I'm adding. Your total is $13.55."

"How do you know?" she demands.

Jim hands her a piece of paper and a pen. "Go on," he tells her. "Add 'em up."

She does: $13.55.

Her son, who is standing right there, says: "See, Mom, the man was right." She glares at him, but he continues chattily, "I could have told you that. I've been in here before and seen him do it. He's always right."

"So I guess everybody is mad at me," John says. We're sitting in the bar late the next morning drinking coffee. He's come to watch the store while his sister takes his father to the doctor. "Margaret's mad at me for not opening up. And now Dad's foot is messed up again." He sighs. We talk for a few more minutes and then the front door opens and his sister and father walk in.

John gets up reluctantly and goes into the kitchen. I hear him and Margaret talking in low voices until John says loudly, "I'm sorry. Look, I'm sorry."

I start to leave and run into Margaret in the store. She shakes her head. "I told everyone this would happen. The doctor said, 'Close the store. Because if he doesn't stay off this foot, by Friday he's going to be back in the hospital.'" Her voice cracks and the tears aren't far behind.

"So we'll close it," she says. "Because I can't be here at 4:30 A.M. and still work until 2 A.M." Margaret is looking at me, but she's really speaking to John, who hangs in the doorway uncertainly.

That night, Jim takes a painkiller and oversleeps. For the first time in its history, the store doesn't open on time. Clemson, a TV sportscaster and regular patron of the store, comes down to drop his wife off for the early train and finds Guinan's front door locked. Clemson then calls John's arborist boss, Lew, who calls John. John rings his father, wakes him up and then rushes down to the store himself.

By the time he gets there, customers are inside, fixing their own coffee and helping carry in the newspapers while Jim groggily sets out the change. "Hey, we got it under control, John," everyone jokes, slapping him on the back. His father doesn't say a word.

The next morning John is back behind the counter at 5 A.M. When I walk in a few hours later, he's wrapping the doughnuts and bagels as if he never left.

I look at him questioningly. "Hey," he tells me. "You do what you've got to do."

Just then, old Ken Anderson strolls through the door, his khaki pants folded up his calves even farther than usual. The weatherman is predicting a heat wave with temperatures jetting over 100 degrees.

"It's a beautiful morning," Ken says to both of us, reaching for his paper, and then adds after a pause, "It's going to be a horrible day."

"You expecting rain, Ken?" John asks.

"Why?" Ken says.

"Because your pants are rolled up to your goddamn knees," John says, smiling for the first time since I arrived.

"It's hot outside," Ken replies. "I gotta keep the air circulating."

⌒

JUST BEFORE MY grandfather died, I found my mother crying in our kitchen. What's wrong? I asked, fearing the worst. Are you sick?

She wiped her eyes with a dish towel. "No, no," she said, searching my face, deciding something. Finally she sighed. "I'm not sure you can understand this. . . ."

Yes I can.

Another sigh. "Okay, well, you know how your daddy tells you he loves you?"

I nodded.

"Well, your grandfather isn't like that with me," she said. "I know he loves me, but he just can't tell me how he feels."

At the time I tried to understand, wanted my mother to think of me as an adult. But I was still too young to believe that my grandfather might have faults. That he was not as I saw him: perfect.

But he lets you kiss him on the cheek, I told her.

"It's not the same—he doesn't kiss me or tell me he loves me," she said, almost angrily now. "Sometimes I just want to hear him say it. But I don't think he ever will."

the general

He took care of the people.

As the mornings passed and conversation flowed behind the counter at Guinan's, it became clearer to me why John and Margaret wouldn't, or perhaps couldn't, let go of the store. Their father had retired for all intents and purposes, but because he lived upstairs and because he didn't own his house, it wasn't as if he could just throw a party, empty the coolers and be done with it. There was the matter of rent to pay, and a lease to negotiate.

At the turn of the twenty-first century, only a handful of homes along Garrison's Landing remained in private hands. The majority of the property was divided between two groups. One was the not-for-profit Garrison's Landing Association, which maintained the gazebo, park and tiny stone theater that long ago had served as the town's train station. The other entity was Garrison Station Plaza, a for-profit group comprised primarily of stockholders who'd inherited their shares over the last few decades. This was the entity that owned the buildings rented by Jim Guinan, David the Australian bookman, Georgia the masseuse and many other businesses operating along the landing.

Despite being "for profit," rent rates for the Station Plaza, including Jim's $1,200 a month, lingered below market rates, and no dividends were paid to Station Plaza shareholders. What's more, the largest voting stake belonged to the nonprofit Garrison's Landing Association. So the only real means for Station Plaza shareholders to benefit tangibly from their holdings was to donate shares to their sister group for a tax write-off. Many holders simply kept the old-fashioned certificates tucked away as slivers of local memorabilia.

There had been a long-standing balance in Philipstown between the haves and have-nots that lent the area a kind of understated grace. But, as in so many places across America, certain forces threatened to tip the balance. The Hudson Highlands, with its bucolic scenery and proximity to New York City, had been "discovered" as a desirable weekend getaway spot. While everything maintained its old-time, dressed-down feeling, the new money had helped trigger tiny shifts in the town's cultural tectonic plates. Upscale local restaurants could command New York City prices. One week a Help Wanted ad appeared for a "Latte Artist" to sell Starbucks brand coffee at the Cold Spring Depot. And with home prices on the rise, the working class more and more was pushed north to seek affordable living.

As part of this shifting, there were—as the rumors suggested— one or two Station Plaza shareholders who wouldn't have minded replacing Guinan's with something more upscale than a dog-eared country store and beer bar. Something that might provide a more lucrative return on their investment one day. While their opinion was not the majority one, they made their voices heard in the whispering asides that then got passed around town and came back to the Guinan family's ears.

Because of this, the family's relationship with their Station Plaza landlords was a complicated one. On the one hand, the Guinans resented the notion that anyone might want them gone after their years of service to the town. Yet at the same time, the store owed its very existence to the Station Plaza, whose early founders had conspired in the late 1960s to protect Garrison's historic river landing from the real estate wheeling and dealing they foresaw sweeping America.

Thanks to those founders—and in particular the friendship of one wealthy man—Jim Guinan had survived rocky times that might have forced him to close long ago.

"HIYA, MR. OSBORN," Jim Guinan called as he walked toward the marina.

"Hi, Jimmy." The General waved back from his car.

It was 1957, and Jim, Peg and the children had recently arrived in Garrison. In addition to his construction jobs, Jim was helping to build the local marina. The project was part of a larger community effort to resurrect Garrison's Landing and its properties, which by the late 1940s and '50s had fallen into disrepair from the economic turmoil of the Great Depression.

Heading up the charge, at a formidable six feet ten inches in his prime, was General Frederick Henry Osborn—or just "The General," as most folks called him out of respect for his designation as major general during World War II. The General was a descendant of one of Garrison's wealthiest and most elite families. His grandfather, William Henry Osborn, was the very railroad tycoon who'd constructed the tall castle looming over Garrison, with the help of popular artist Frederic Church, whose works hang in the Metropolitan Museum of Art and in other museums. Osborn was one of many illustrious residents whose sizable wealth helped earn the Hudson Highlands the designation of "Millionaires' Row." Others included Hamilton Fish, former New York governor and secretary of state, and noted architect Richard Upjohn. Banking tycoon J. P. Morgan later bought his own lavish estate across the river from Garrison.

It's easy to understand the attraction of a place like the

Highlands for these men. The economy of the mid to late 1800s was marked by tumultuous periods during which there would be boom times followed by panics and depressions. By contrast, the bucolic region around the Highlands offered corporate chieftains a calm respite from their bare-knuckle business dealings. Moreover, disease was rampant in the urban streets back then. Cholera struck New York City several times in the second half of the century. There were outbreaks of yellow fever, tuberculosis and malaria. To keep a home in the Highlands, amid the fresh breezes of the mountains and forests, was no insignificant luxury.

After William Henry set down roots, the legacy of the Osborn clan loomed large through the region. This was not only because of the family's vast landholdings and wealth, but also because of their growing push to preserve the Hudson Highlands for the next generation. For although the railroads and accompanying Industrial Revolution brought the clan great wealth, it also wreaked havoc on the environment, given the fledgling country's insatiable appetite for timber to build the tracks and fuel the locomotives. And the Highlands, with its swaths of trees and proximity to the river for shipping, had proven a fruitful provider.

By the end of the nineteenth century, faced with the ongoing wreckage of the natural beauty surrounding their country homes, many captains of industry began to turn their well-endowed energy toward conservation. Painters and writers had all but ignored the deforestation in their earliest romantic Hudson River works; but they too began to condemn the destruction. Over the course of several decades, public outcry intensified and resulted in millions being invested to preserve the Hudson Valley.

The mission eventually passed into the hands of William Henry Osborn's heirs. Along with protecting the natural habitat came a new zeal to resurrect and maintain the sites so critical during the nation's early years—including Garrison's own dilapidated little river landing.

And so it was that Osborn's illustrious grandson, the General, came to befriend a humble Irishman named Jim Guinan.

⌐———◦

TO LIVE ON the landing in the 1950s and 1960s as the Guinans did was to be literally from the wrong side of the tracks. Yet both the General and his gentle wife, Margaret, were frequent faces there in those days as they pushed to revitalize the landing. When the General's father, William Church, died, he left his daughter-in-law Margaret a gift of money. She used the greater portion to help fund the construction of a marina down at Garrison's Landing with an adjoining waterfront park. When Jim Guinan arrived on the scene in 1957, he joined in the construction.

The General lived up to his nickname in that he was a natural leader whose chiseled features and sonorous voice made him able to organize and dictate the world around him according to his design. He had built his own Garrison empire on another mountaintop—it was appropriately dubbed the Big House as its doors, window views and details were designed to accommodate his own sizable proportions. The General was a lean man who sharpened his own axes on a grinding wheel in his study and felled the trees along his drive that might shade the pink mountain laurel. One winter Saturday, after a storm severed a power line between the main road and his property, Osborn—by then in his seventies—descended over a cliff to retrieve the line, splice it and restore electricity.

As a younger man Osborn didn't hesitate to roll up his sleeves and plow in with Jim and the other laborers to help with the marina's construction. He was such a presence that Jim and his crew eventually named their pile driver the "General Osborn," and despite their different social strata, Jim felt free to call out to the General each time he drove by.

"Hiya, Mr. Osborn."

"Hi, Jimmy." The General would wave back.

Margaret Osborn, meantime, was an artist and gardener who used her talents to assist with the plantings along the riverfront. One day she was coming up from the riverbank about the same time as Jim's wife, Peg, was tossing out a pail of ashes from the stove into her rose garden—the contents of which landed all over Mrs. Osborn.

Horrified and eager to make amends, Peg insisted the lady come in for a cup of tea and to clean up. They chatted about having the same name (Peg being an Irish nickname for Margaret), shared gardening tips and, as they sipped tea into the afternoon, found they liked each other quite a bit.

TWO YEARS AFTER the Guinans arrived, the old Belcher house at the end of the landing became available when the aging proprietors decided to retire and move. Since its construction by Belcher in the mid-1800s, the building had almost always doubled as a home and business, providing some service to the community, from meat market to paint store and then a grocery under the Yannitellis and eventually its current incarnation as a general store and pub. Worried about the possibility of the historic property falling into the wrong hands, the General approached Jim Guinan about moving into the house with his family and taking over operation of the enterprise.

"You have a family, this would be good for you," Osborn said in his typical straightforward manner. "And you'd be well liked around here."

The General then offered to sponsor Jim to buy the house, leaving the $242.82 mortgage payments to the Guinans. On June 30, 1959, Jim and Peg fastened their names to the deed transferring the store and bar into their name with the General's backing. After a lifetime of squeezing into tight spaces, the large frame house was big enough for each Guinan to finally have his or her own bedroom. Jim and Peg's quarters faced west, and from their bed the couple could watch the river and lights of West Point as they drifted off to sleep. Downstairs, there was room enough for a dining table, couch and pets with names like Big Foot and Sergeant, who trailed in and out with the customers.

Almost from the beginning, though, the Guinans had difficulty keeping up with the mortgage. For starters, there were other general stores and bars around providing competition. What's more, times were lean for the working class in town, and the Guinans were often generous with their payment terms. At Christmas, Jim and Peg would set out toys in the store for parents to buy, and then allow repayment at a rate of one dollar each week. They also began keeping a "good book" of their own with customer accounts. Inside its pages, they continued a tradition practiced by the small shops in Birr of letting folks charge what they needed and pay it back when they could, which wasn't always as promptly as hoped. Meantime, their bills to vendors and for the utilities stacked up.

"I couldn't keep up," Jim recalls. "It was very hard. I was buying the groceries. There were two bread people. Three meats. You had to have money up front for every one of them."

As the bank threatened foreclosure, Jim went to the General and explained his dilemma. The General listened and, as was his way, immediately proposed a solution. "Sell the house back to me. You can pay me rent and continue to live there and run the store." So on the fifteenth day of April 1961, less than two years after they'd moved in, Jim and Peg signed the deed to their home over to General Osborn. And with this deal, the business stayed alive and soon became a cornerstone of the community.

The official hours were 4:30 A.M. until 11 P.M., but that was just what was posted on the door. In truth, Guinan's was open twenty-four hours a day, seven days a week, 365 days a year. Some nights Jim would awake to a rap on the door from the members of the local boating club who wanted a few drinks following one of their meetings. The Irishman would unlock the doors, pull out beers and sit with them until just before dawn. At that point there was no sense in returning to bed, so he'd just stay up and wait for the sound of the newspaper deliveryman on his doorstep. Guinan's also became the feeding post for the local volunteer fire department, which operated via a relay telephone hotline. Soon as she got a call, Peg would come down in her bathrobe to start making sandwiches and coffee, and shortly afterward Barbara Prescott would walk over and pick up the food, cigarettes and soda to carry to the crew.

People in town from different backgrounds and economic means came to feel as if Guinan's was their home too. During the racial tensions of the 1960s, Jim befriended a black highway worker by the name of William "Buster" Coleman. Worried that he would be robbed, Buster entrusted all his savings to Jim to keep locked up at the store. Buster drank beer with the white men at the bar from day one, and it was never an issue because Jim and Peg refused to let it be one. On sunny days, Buster would sit on the front porch telling ghost stories to the Guinan children while he dangled a paper bag filled with pebbles over the railing to rattle at the scariest parts. Years later, when Buster was in the hospital dying after an accident on the job left him paralyzed, Jim and Peg turned over their bedroom to Buster's sister and niece from Florida when they came to visit him. When things looked grave for his friend, Jim came to Buster's hospital bed and delivered his four thousand dollars in savings—all in cash. "I gave it back the way I got it," Jim says.

Despite the time and energy put into the store, even the now lower monthly rent payments to the General were still something of a struggle for Jim and Peg. In December of 1962, a strike closed down major New York City newspapers for 114 days and in the process stripped the Guinans of a steady source of income as well as one of their most trusted customer draws.

Each year, harsh northern winters sometimes kept shoppers at home, yet the Guinans still had to pay their overhead of heat, inventory and electricity.

But the General was far more lenient than the bank and trusted Jim to pay when he could. Managing the General's landing properties was Agnes Donohoe Preusser—my friend Ed's headstrong grandmother, who dominated the real estate landscape back then, managing properties for the town's elite.

In a typed letter dated March 31, 1964, Agnes wrote to the General at his Park Avenue apartment in New York City.

"Dear General Osborn . . . Mrs. Guinan tells me, she will write you to state that they will pay the back rent they owe in a week or ten days. The winter oil bill has been high. They were hoping you might stop in or they would have written you before. They tell me all their bills are paid except this rent they owe you. Jim is working steady [on construction]. . . . I am sure they will send you a check very soon. I know they always keep their word."

And so, against the odds, and with the generosity of the General, Guinan's kept its doors open.

BY THE LATE 1960s, 20th Century Fox was searching for a spot to build the elaborate sets for its movie musical *Hello, Dolly!* starring Barbra Streisand and Walter Matthau. The setting was to be 1890s Yonkers, but the problem was that Yonkers in 1968 was the state's fourth largest city with a population of roughly 210,000, and didn't exactly project the feel director Gene Kelly and others wanted.

Garrison's Landing, on the other hand, was perfectly undernourished and, thanks to the preservation efforts, tailor-made for such a set with its clapboard dwellings and old shops. The river landing had a population of about sixty, which made for a manageable set.

While the film's economic infusion promised to be a boon to the run-down landing, it also worried longtime residents—including the General—that Garrison's newfound prominence after the movie's release would pique the interest of real estate buyers who cared little about preserving the town's history and homey sensibility. And so the landing's major property owners came together and devised a strategy that would protect the strip long after 20th Century Fox packed up and left.

When the details were complete, the General returned to Jim Guinan again. There was a plan, Osborn explained, one he hoped would prevent future developers from coming in and destroying the historic feel of the area. He told Jim not to worry, and that rents would be kept reasonable to help retain hardworking tenants such as the Guinan family.

After the two men finished talking they stood—Osborn towering over the small Irishman—and shook hands firmly.

"See ya, Mr. Osborn."

"See you, Jimmy."

DETAILS OF THE plan were publicly disclosed on January 24, 1968, several months before a movie construction crew descended on the landing to prepare it for immortality on the silver screen. The General's wife, Margaret, had prepared a speech for a gathering of townsfolk. Winter was firmly entrenched, and the landing's trees were barren in the cold. Margaret kept her remarks to five and a quarter typed pages, double-spaced.

"Twelve years ago, my father-in-law William Church Osborn died. He was a most public spirited citizen and was devoted to the welfare of Garrison, the Hudson River Valley, New York City, New York State and our country. It was in his memory that a few members of his family planned a small memorial park opposite the station. . . . After the park was completed, willow trees planted, grass sowed, paths laid out, hedge and shrubs planted, the idea grew to have a marina in connection with the park . . . something Mr. William Church Osborn had always been keen about. . . ."

As she continued, Margaret Osborn laid out a vision for the future of Garrison's Landing and presented a business road map for its preservation, much as her father-in-law, and his railroad magnate father, would have wanted it. Henceforth, the marina, park and theater would be overseen by the not-for-profit Garrison's Landing Association, which had been given an endowment fund. To complete the picture, she told the assembled crowd, a group of local men had stepped forward to form a real estate company called Garrison Station Plaza. The Plaza now owned most of the remaining waterfront buildings and would keep them in shape. What's more, the lady added, substantial blocks of Garrison Station Plaza stock had been given to the nonprofit Garrison's Landing Association.

The purpose of this final transaction was well reasoned, she noted, "So the two are closely related and we hope that the area will always retain its old-time flavor and charm."

A FEW MONTHS later the movie crews descended to work their magic. They spent sixty days overhauling Garrison's Landing, and by the time the transformation was complete, some $1.5 million had been spent, employing 7,500 pieces of lumber, 1,000 gallons of paint and 100,000 square feet of sod to landscape the riverfront park. Garrison's brick Coal and Oil Company building—once the hotel and brothel—became headquarters for the fictitious Vandergelder's hay and feed store. The old railroad station was refurbished and the streets macadamized so they could be temporarily transformed into 1890's red brick. To achieve perfect outdoor color, two workmen labored for hours stripping leaves from a tree in the waterside park and replacing them with stapled-on artificial foilage imported from Italy. With an estimated twenty-million-dollar budget, *Hello, Dolly!* was at the time the most expensive musical ever filmed.

The film's impending arrival rippled through town. Ed and John Preusser's mother, Nora, just out of college and freshly married to Ed senior, a young corporate pilot, was hired by 20th Century Fox to be Barbra Streisand's personal assistant during the film. The first morning Nora met the star, she recalls, Streisand walked out of her bedroom, no makeup, in a night-gown, twirling her very small, scorched silk bikini underpants and wondering who had put them in the dryer. Nora liked her immediately. The young Garrison girl took Streisand's son, Jason, eighteen months old at the time, to buy his first pair of "hard" shoes and arranged for any specialty food, such as lobster, to be delivered upon Streisand's request. Paid twenty dollars a day whether she worked or not, Nora could come and go freely on the movie set and into Streisand's dressing room, where she advised her boss on where to shop for antiques and get the best deals. Streisand was generous in return, giving Nora steak knives, albums, signed photos and—better yet—keeping in touch over the years when she was interested in looking at real estate property in the area.

The General, meantime, held meetings at the Big House, hosting Streisand and Matthau as well as director Gene Kelly for dinners and hammering out financial details with 20th Century Fox executives. His grandson Fred Osborn III recalls once watching the General from the doorway as he negotiated a deal at his dining room table. "At one point, all the chatter stopped, and my grandfather just took over. That was his way."

Jim Guinan conducted his own wheeling and dealing down on the landing. When movie executives told him that access to his store would be shut off and that he might want to consider closing down during filming, the Irishman got huffy. "I'm telling you, it took me ten years to build this business, and I haven't closed down once," he responded indignantly. "I'm not going to lose customers for you to be here for a few weeks." Instead, he arranged to get food deliveries brought across the railroad tracks, and his loyal patrons took the same route themselves. "You couldn't stop them from coming, not my customers," Jim notes proudly. After shooting officially commenced, Guinan's quickly became command central for the film's crew, with Gene Kelly, the actors and workers all stopping in regularly for a bite or beer.

The kids loved the action from the start. John Guinan remembers throwing a Frisbee to Streisand and being impressed when she caught it and tossed it right back. "She had good aim," John notes. One day his brother, Jimmy, rode his bike up to a bunch of parked stretch limousines and chatted away with 20th Century Fox executives about the local lore as if he himself were mayor. A few days later one of the men strode into Guinan's and announced he was looking for a certain "Jimmy Guinan." Peg was behind the counter and called quickly to her husband. "No," the executive corrected. "The kid."

Impressed by Jimmy's gumption, the man wanted the youngster as a guide and to help take care of the movie set's chickens and rabbits. In return, Jimmy negotiated bit parts in the film for himself and his dog, Sergeant. He also insisted on roles for his poorer pals living down on the landing who had been passed over in favor of children from wealthier families.

Some of the adults grumbled to the media about the uproar— "I think it looks like hell, all them signs," griped thirty-two-year resident George Dale to the New York *Daily News*. But even they grew accustomed to the hubbub, and the production crew took pains to be considerate. Every morning they would take TV antennas off the homes for filming and put them back up before leaving for the evening. When filming was over ninety days later,

20th Century Fox bought new ones for everyone to make up for the wear and tear.

That winter, with the *Hello, Dolly!* set torn down and Barbra Streisand and Walter Matthau gone from Garrison's Landing, General Osborn signed the deed passing Jim's home into the holdings of Garrison Station Plaza Inc. It was December 18, 1968, and almost a decade had passed since the Guinans had set foot on America's shores searching for work and a better life.

The General and the Irishman stayed on good terms, with Jim and Peg attending the General's fiftieth and sixtieth wedding anniversary parties. Jim was also at the General's funeral when he finally died on January 5, 1981, by which time he was ninety-one and his height had diminished to a still formidable six foot eight inches. During the service, Jim Guinan stood quietly near the door in the back of Garrison's stately St. Philip's Church and bid a silent farewell to the unlikely friend and mentor who'd forever changed his life.

A year later, the General's wife died; they are both buried around back of the church in the vicinity of an enormous white cross that marks the territory of all the Osborn family plots. Engraved on the couple's marker are cardinals, flowers and two ducks facing each other with three offspring trailing behind each one. There are also these words:

> *The Road to Heaven is Faith*
> *The Road to Heaven is Love,*
> *The Road to Heaven is Living and Giving,*
> *The Road to Heaven is Heaven.*
> —*Frederick Osborn*

"He was a helluva nice man altogether," says Jim. "Really, he was. He took care of the people."

Slainte go saol agat,
Bean ar do mhian agat.
Leanbh gach blian agat,
Is solas na bhflaitheas tareis antsail seo agat.

Health for life to you,
A wife of your choice to you,
Land without rent to you,
A child every year to you,
And the light of Heaven after this world for you.

——*Irish toast*

14

nectar of the gods

Boys, here's to Cliff.

Usually Jim sticks to nonalcoholic Haake-Beck. But sometimes, if his kids aren't around, the barman will pull a regular
light beer from the cooler and sip it with a few folks in the bar.
"What the hell," he'll mutter. "People smoked and drank a little
nip of whiskey forever and lived to be a hundred. My mother-in-
law was in the nursing home and she had a prescription for a shot
with her dinner. That's the way they done it then. And she was
all right."

I never see him drink more than one, but that's one more than
Margaret will tolerate. When she finds an empty can in the living room one afternoon and then later a bottle in the kitchen,
she smashes her hand in frustration against the glass pharmacy
dispenser, rattling the old Aqua Velva and Vaseline Hair Tonic
bottles inside. "You aren't supposed to have the carbohydrates,"
she yells, her eyes reddening. "Do you want to lose the foot this
time? Is that what you want?" After that, when John finds an
empty beer can lying around, he confronts his father—"Damn
it, would you at least hide the evidence, Dad?"—but throws it
away before his sister arrives.

FOR A BARMAN, booze is something akin to a workplace benefit, or hazard, depending on your point of view. Jim Guinan had always maintained a natural affinity for the drink—"NECTAR OF THE GODS," he'd call out cheerfully back when he regularly drank beer, particularly Rheingold.

In the old days, to drink at Guinan's was a young man's coming of age, for the cadets, the workers' boys in town, even for Jim's own sons. When Jimmy turned nineteen, he brought home a flock of friends on New Year's Eve and left them in the bar while he went upstairs to shower. When he returned, his pals were seated at the bar drinking a specially brought out bottle of Paddy's Irish Whiskey with his father, faces proud and backs straight. "My friends and I, we were elevated somehow," Jimmy says. "It was a big thing. My dad started calling me Buddy right around that time."

As long as Peg was alive, Jim kept his alcohol consumption in check. His wife never drank—would walk around a party holding the same untouched glass of wine for hours to be polite—and while Jim might have been the front man, everyone in town knew that Peg was the boss. Every morning, she cooked a hot breakfast for her kids—scrambled eggs, porridge, fried tomatoes, French toast, bacon—refusing to let them out the door until they'd sat down and eaten. Meantime, she helped tend to the store, balanced the books, did the laundry and pumped gas down at the docks. Customers grew accustomed to her thick accent and hearty laugh; there was always a cup of tea or plate of biscuits ready to go for visitors. Bartending was left to Jim, but he even took his wife's cue in there when it came to rowdy behavior. If a bloke cursed, Jim would tilt his head and glare over his eyebrows as fair warning just as Peg's long shadow loomed in the doorway to warn: "One more, the door." Her children were taught to address customers as Mr. Dain, Mr. Cuneo, Mrs.

Travis and forbidden to call each other "pigs" or "stupid" or to say "What?" when called. "You should always say 'Yes?' " Peg drilled into them.

Peg was such a force that no one really noticed she was feeling bad in the spring of 1986, mostly because she never slowed down enough to let anyone know. A hernia from lifting boxes of canned goods finally drove her into the hospital for surgery, where doctors performing a scan of her GI tract noticed an inflammation around her pancreas. She was also jaundiced, which indicated that something was amiss with her liver. So the children took her to Westchester Medical Center, where doctors did biopsies and found a tumor in her bile duct.

They treated everything the best they could, and yet for the next year, Peg went steadily downhill. Bouts of pneumonia, congestive heart failure and her hernia all conspired with the bile duct cancer to steal her strength until finally it was a triumph if Margaret, who was taking care of her mother in between stints at the precinct, could get Peg to walk to the mailbox and back. Barbara Prescott would come down and sit with her old friend, bringing her an apple juice and filling in behind the counter when the store became too much for Jim.

On Easter Sunday morning in 1988, Peg's daughter Christine, who was living nearby, went into labor with her third child. When her water broke that morning, Christine's first instinct was to go find Mom. She arrived at the store and found her sister Margaret preparing Easter supper and overtly distraught. Their mom had suffered through a rough night, Margaret said. She'd been running upstairs to her parents' bedroom all morning checking on Peg, and things seemed to be getting progressively worse. But there was nothing to be done at that very moment because Christine needed to get to the hospital.

After dropping her sister and brother-in-law off, Margaret returned to the store and found her mother comatose. Her dad was out in the store taking care of all the holiday customers stopping in to bid them Happy Easter. What happened next is sketchy in everyone's memories, a side effect of the day's numbing gravity. An ambulance was called; John arrived at the store after Sunday Mass. And in the confusion, while emergency workers carried Peg's frail frame downstairs, Margaret took her brother aside in the kitchen. "Listen to me," she told him, "Mom has spoken to me. She wants no extraordinary—"

"I know," John interrupted.

"No, you don't," Margaret said.

"Yes, I do," John said. "She spoke to me too. No extraordinary means." He hugged his sister and followed the stretcher out the door.

Just after noon, as the ambulance was leaving Garrison's Landing carrying her mother, Christine gave birth to her son James. When her husband, Mike, called the store with the news, he learned Peg was on her way there with John. As soon as Jim and Margaret could square things away at the store they'd be along.

But time wasn't on their side. Doctors at the hospital quickly pronounced Peg's situation grave. A nurse who was a friend of the family's called Margaret. "You need to come here now," she said. So Margaret telephoned her boss at headquarters and told him she needed a personal day. She shut off the oven, leaving their Easter turkey and ham half cooked, and then hurried her father to the car.

Meantime, in his father's absence, John relayed his mother's wishes as she'd laid them out for him and Margaret. "She has a do-not-resuscitate order," John told the doctor. "She's had a long wonderful life, and if it's time for her to go, it's time for her to go."

The nurse brought young James in from Christine's room and laid him on Peg's chest in the emergency room. A hush fell, and for a moment, Peg's shallow breathing seemed to even out. The newest Guinan clenched his tiny fists and squirmed against his grandmother for the first and last time. Afterward, they moved Peg to a private room to rest, and to wait.

By the time Jim and Margaret got to the hospital, Peg Guinan had died. Jim asked his children for a moment alone with his wife. He went into his Peg's room and closed the door to say good-bye to the beautiful girl he'd spotted working down the line from him nearly four decades before. When he came out of her room, he was crying for the first time in John's memory.

Margaret didn't stay at the hospital long. There were details to tend to after all. She left and drove straight back to the store. First, she found a bottle of scotch and poured herself a good stiff drink. That day was the end of one path in her life and start of another. From then on, Margaret would have to fill her mom's shoes. She would be business partner, nurse, cook and daughter to her father, and at times she would be mother to the rest of her siblings. Some days she would do it because she wanted to, some days because she had to and others because she didn't know how not to.

The duties started immediately. After a few sips, the detective pushed the scotch bottle aside, picked up the telephone, and started making phone calls.

The Guinan daughters are superstitious. Soon after her mother died, Margaret, who was still living at home and taking care of her father, had a vision of Peg walking toward her, a sign she believed was her mother telling her not to worry, that Margaret was doing a good job carrying on in her absence. Even now, the daughters still look for guidance from their mother. If they lose something, they ask her for help to find it. If one of them stubs a toe, it's surely Peg trying to tell them something. And Christine knows that there is no coincidence about the day of her mother's death.

"Mom always told us that in order for God to make room for a new life, another must go," Christine says. "She waited for James."

THE FAMILY WAKED for two days through a steady procession of well-wishers from the town and counties all around. The store closed for the wake, then opened back up immediately. For months, Jim was never alone. Margaret was still living at home. Plenty of folks dropped by regularly to pay their respects. Jim couldn't have a cup of tea in the morning without someone peeking in to check on him.

Eventually, though, people started to move on with their lives. People still talked about Peg, but not as often. In 1991, Margaret finally moved out of the house as she moved up in the ranks of the police force. John was working at General Motors down the river and helping out occasionally at the store. Jimmy was in California. So there was just Jim to keep things going, and he tried, along with Barbara Prescott, who began to help out regularly in the mornings. At first she worked 8 A.M. to 3 P.M. and Jim opened up. But that got to be too much for him after staying up to close the night before. So Barbara took over the early shift, even though she was beginning to see signs of trouble with her feet from diabetes complications. Murray came with her to carry in the heavy stacks of newspapers before heading up to the local school where he worked as a jack-of-all-trades.

And slowly change snuck up on Guinan's.

If Jim was lonely, he never said so to his children. Nor did he ever lament to his customers, instead keeping up the same cheerful—"How are ya," "Good to see ya"—rhythmic banter he'd perfected over the years. But soon after Margaret left, he invited a fellow named Cliff who'd gotten kicked out of his own place next door to live upstairs with him. In exchange for room and board, Cliff worked behind the counter and kept Jim company.

In late 1993, with Cliff still there, Christine and Mike came in to run things at Jim's suggestion. In many ways, they carried on Christine's parents' old traditions—writing up the Reserved stacks, keeping the old Federal cash register. One day a young girl walked in and asked Mike to lend her twenty dollars so she could get to the city and then catch a flight home. With the boys in the bar chuckling, Mike reached in his wallet and gave her the money along with his name and address. That, after all, was how Jim and Peg would have done things, trusting until there wasn't a reason to.

The boys razzed Mike for days. But two weeks later a card arrived at the store with a check for twenty dollars and a note from the girl's parents thanking this stranger who had helped their child and given them renewed hope that "there were still some good people in this world."

Still, Christine and Mike were young and ambitious and wanted to improve the store's fortunes. They made modest modifications—moving coolers to facilitate more foot traffic, adding an electric coffee maker so they didn't have to make each cup by hand. Even this little bit of change seemed to rub Jim the wrong way. He tried to explain that this wasn't the way Peg and he had done things. Mike would alter something and turn to find his frustrated father-in-law standing there, hands on hips, demanding: "Now why are you doing that, lad?" And so, more and more, he and Cliff would get out of the house and pass the afternoon having lunch and a few drinks in the sanctity of some local pubs.

ONE OF JIM's favorite stomping grounds was a place called Jeremiah's in Peekskill. The bar, now under different management, was located at number 50 Welcher Avenue inside a low-slung brick building with a cramped parking lot and low-key signage. Back then, the main decor was frog paraphernalia, a nod to the seventies hit "Joy to the World" celebrating the bullfrog

Jeremiah. The bar later became the NY Firehouse Grille, and axes and flame-retardant suits came to replace the amphibian wares.

Getting there from Garrison's Landing meant taking one of two roads: the winding Goat Path, which whips around hairpin turns off Anthony's Nose mountain, or the less treacherous Route 403 leading to Route 9, which runs parallel with the river up to the exit for Jeremiah's. This was the route preferred by the Guinan's crew, which typically consisted of Cliff, Jim, their friend Stevie and occasionally Fitz or another Guinan's regular.

As soon as someone arrived to relieve them of duty at the store, the boys would pile into the staff car and, with Cliff behind the wheel, set off for lunch. When the Guinan entourage arrived, the bartenders immediately treated them like royalty, and hold court they would. Cliff was a heavyset, curmudgeonly fellow with a peppered beard that made him look like Santa Claus. The softhearted, mustachioed Stevie was the quietest of the trio. He was a schoolmate of the Guinan kids and had known the family most of his life. Both Cliff and Stevie were regulars at Jeremiah's, but Jim—well, he was the king.

"Set 'em up," the Irishman would call out, his accent singsongish with the promise of a fine afternoon to come. *Knock, knock,* he'd rap his knuckles on the bar. "Set 'em up, on me."

Usually it was Dewar's scotch, maybe with a little water. Occasionally, if the mood struck him, Jim would call for a glass of Tullamore Dew, a particularly strong brand of Irish whiskey that he favored. He'd chide anyone who drank the whiskey too quick—"for the luv of God, lad, it's a *sippin'* whiskey." They'd drink, toast, tell stories, flirt a little with the prettiest bartender and buy each other rounds.

THE THING ABOUT alcohol, particularly the sweet stuff, is that it can wreak havoc on blood sugar levels, which diabetics must control tightly to keep infection at bay. What's more, booze is thought to be toxic to nerve tissues, which is in part why doctors encourage patients with diabetes to curtail their drinking. Since one of the longest nerves in the body comes off the spinal cord and branches down into the big toe, heavy drinking created a poisonous environment that ran down to Jim's feet. As the years passed, doctors believe the alcohol coupled with the neuropathy and standing on his feet all day eventually produced the equivalent of a diabetic's perfect storm in Jim. He was already standing off balance because of a lack of sensation in his feet. So a callus developed, but it didn't hurt him because he couldn't feel it. Then the callus became an ulcer and the ulcer got infected and that put him in the hospital.

Jim sees it differently. The infection and the drinking are coincidence, and he can tell you exactly why. For starters, he was always careful when he drank. Would sometimes coat the stomach first by gulping a raw egg cracked into a glass of beer. What's more, he mixed his Dewar's with water and some ice—"The water, it didn't hurt the stomach. I drank Dewar's and never got drunk." And finally, Jim points out, he always made sure to eat plenty. "People forget to eat, see," he explains. "And that's why they get drunk. If ya eat, you'll be fine. As I said before, it's like smoking. People been smoking for years and never died of cancer. Take the bums on the road. They all smoke. They all drink. And they are fine."

Jim pauses his explanation briefly to pop a handful of pills into his mouth from the bottles on his kitchen counter. In the counter's collection are: Glucophage (850 mg), Glucotrol (5 mg), Cozaar (50 mg), Pletal (100 mg)—medicines to control diabetes, high blood pressure and poor circulation. He swallows them quickly with a sip of water, then wipes his mouth and continues his train of thought.

"Drinking—depends on how you treat yourself. You gotta eat. And that's the truth, luv."

THINGS GOT BAD with Cliff. All the boys were drinking, but Cliff, he never seemed to stop: Budweiser in the bar, scotch at lunch, a Pepsi bottle filled with vodka and a splash of the dark soda ("for the color," Jim says) during the times in between. He stashed half gallons of booze under his bed. Margaret tried playing tough with him. She'd open up his hidden bottles and pour out the contents. She took his keys from him, threatening to call her police contacts and have him thrown in jail if he drove. "Cliff, I know where you eat lunch every day. All I've got to do is call in a favor." But she never did, and none of it seemed to faze Cliff.

So the two Guinan sisters took their father aside. Begged him to stop drinking with Cliff. Christine even cried and showed her father an essay her son had written in school that began, "My grandpa likes to drink." Just ask Cliff to leave, they said.

But Jim wouldn't hear of it—"Where's he going to go?"—and it was Jim's home, so Cliff stayed. And Cliff was pretty handy, would try his hand at fixing almost anything—TVs, computers, boats, plumbing, you name it. Even folks who found Cliff gruff and uncooperative when it came to tasks like making sandwiches would call down to Guinan's when they couldn't figure out what the problem with their sink or stereo was. Besides, he was good company, a steady sidekick to Jim. Together they would go across the river to pick up tickets for the football tailgate parties. One Saint Patrick's Day, Cliff sat shotgun next to the bartender as they rode in a parade under the banner "Hudson River Irishman" and drank and danced in the streets well into the evening. Cliff was also with Jim during his 1998 Grand Marshaling in Peekskill; Christine and Margaret discovered their father marching, against doctors' orders, when they turned on the TV and saw him there beaming away.

Eventually Christine and Mike decided they'd had enough of

battling with Jim about Cliff and every decision or alteration they made regarding the store. And Christine could no longer bear to watch her father drink. So in 1997, they handed control of the store back to their father. When they left, Cliff was still there.

Like Jim, Cliff also was a diabetic. But he didn't have an entire town always looking out for him. Eventually Cliff started limping a lot, and Jim noticed bloody footprints on the stairs and bathroom floor upstairs as well as empty hydrogen peroxide bottles in the trash cans. He pushed Cliff to see a doctor, but Cliff insisted he was just limping because he had a bum knee. Then one day they were both sitting in the bar with Old Mike, Fitz and another fellow named Paul. And Fitz started grousing about the smell in the room.

"Jesus Christ, Jim," Fitz complained. "What the hell died in here? It smells like a fucking dead raccoon is laying under the bar."

Jim didn't answer, and a few moments later, Cliff slipped out from behind the bar and limped upstairs without saying a word.

"Look now," Jim said. "There's something wrong with Cliffy. His foot is in bad shape. And he won't do anything."

"Where is he?" Mike asked.

"Back up in his room," Jim said.

So Old Mike took charge. Headed through the kitchen and into the living room and up the flight of stairs to the bedrooms with Paul on his tail. Mike doesn't have a sense of smell, so he didn't stop. But Paul couldn't take the odor, and turned around and went back to the bar. Meantime, Fitz left. Upstairs, Mike walked into Cliff's room and found him sitting on the side of the bed with his shoe and sock off. The problem foot was black and brown with a hole in the bottom and one toe was dangling sickeningly, so decayed it looked like it might fall off.

"Ahh, man, we gotta take you to the emergency room, Cliffy," Mike said immediately.

"I ain't going," Cliff mumbled, groggy from the infection.

"Yes, you are," Mike said. And he went back downstairs and called for help. Soon after, David Lilburne, the bookman, showed up driving Garrison's emergency volunteer ambulance. By then, Jim and Mike had cleaned up Cliff's foot—he was too weak to protest—and bandaged it as best they could. All the boys helped carry his sizable frame down the stairs and then loaded him into the ambulance. Then they stood and watched as David pulled out of the Landing and headed south toward the Peekskill emergency room.

The next day, after doctors amputated his toe, Cliff woke up and recognized Jim sitting in the room with him. "Thank you," he said. "Thank you very much."

"You son of a gun," Jim said. "I hope this is enough to teach you a lesson."

"It is, Jimmy," he promised. "It is."

IT WASN'T. AFTER a short break, Cliff returned to his drinking, though he tried to hide it more. He'd stay up in his room, drinking straight booze and watching videos on the VCR. One morning, he was coming down the stairs and saw his reflection in a mirror and was so disoriented that he startled himself and went plunging to the bottom.

In May of 2001, ten years after he came to live at Guinan's, Cliff drove south to visit an old girlfriend of his and checked into a nearby motel. One day his ex got a telephone call, Jim says. "I'm not feeling that well," Cliff told her. When she went to see him, the motel door was locked and no one answered. The clerk called the police, and when they got the door open, Cliff was lying on the floor dead of a heart attack.

His remains were brought back to New York for a funeral service. Jim attended. So did Fitz and a few others from the bar, but altogether it was a small group. When the tiny Guinan's contingent arrived at the funeral home, they found a few of Cliff's family members milling in front of the urn holding Cliff's ashes, including a daughter Jim had never met. A few folks spoke a bit awkwardly. And so Jim stood up and cleared his throat.

"Ladies and gentlemen, my name is Jim Guinan," he said to the room of strangers. "Some of you may know me and some of you don't. This is a sad day. Cliff, he stayed with me for ten years. He was a great guy and a smart man and a lot of people didn't realize that. Computers, mechanics, TV, plumbing, anything in the world, he could do it all, I tell ya. He was a good man, and he had natural gifts. Why, the phone would ring in my place and it would be for Cliff. People didn't realize it, but they needed him.

"And," Jim continued, "he was my friend."

Jim kept going for a while, talking as if Cliff were the best man who had ever lived, and by the time he was done, several members of the family were crying. Jim paused for a moment and then looked up at the ceiling. "Cliffy, I'm sorry, but this is the one time I wasn't around to answer your call. You'll be missed."

And then he walked over to Cliff's daughter and handed her an Irish mass card he'd brought. "This will help your dad into heaven."

Says Fitz, who stood to the side watching the speech: "Truly, that was one of the finest things I've ever seen Jimmy do. He rose to the occasion when nobody else did."

AFTER THE FUNERAL, Jim and Fitz and a few others drove straight back to Guinan's. When they arrived, there were some folks who had known Cliff sitting back in the pub having a beer and chatting. Jim marched in, took one look around and shook his head in disgust: "You buggers here don't have a lick of decency," he told them, pointing his finger down the line at each one. "Cliffy, he may have liked his drink, but he was there at everyone's beck and call anytime you needed him. And not one of you had the courtesy or decency to come to his funeral."

In the embarrassed silence that followed, Jim strode behind his bar, pulled a beer for himself and raised it high in the air. He turned to Fitz and the few others who'd attended the service.

"Boys, here's to Cliff," he said.

Fitz and the others raised their beers. "Here's to Cliff."

THE BOYS' CHARIOT, the white Volvo, is still parked outside the store.

John tells me its engine is still strong and raring to go. But the tires are flat and rims rusted out, which means the staff car is stuck right where it is.

IN THE YEARS after my grandfather's death, my grandmother slowed dramatically. She put up a good front for a while, but things began to slip.

"Where are you living these days, Wendy?" she asked one day when I was home visiting from New York.

New York, I told her hurriedly. You know that, Grandma. She reddened, and I realized that she didn't know anymore.

The doctors suspected Alzheimer's. And this was just the beginning, these tiny holes in her memory that would soon spread to spill the milestones of a life well lived. Certain things she fought to keep longer than the rest. My grandfather's photo from Pettiford was the one she remembered and wanted to kiss. My birth date, the twenty-ninth, was the number she spoke out of the blue the week I turned thirty-three. And my aunt's voice made her smile even when she could no longer walk, feed herself or sit up straight.

Only later did it occur to me that these milestones that had defined her life most vividly had nothing to do with money, titles or power at all.

the sarge

Hey, *baby,* baby, *can I ask you a question?*

Have you ever been robbed? I ask Jim.

He nods. "Three times—one of them someone broke in the back door and stole all the cigarettes. There were some close calls too. Like the night a bunch of drunk boaters from Yonkers came banging on the door at 4 A.M. demanding fuel. I told them I'd give them just enough to get back to Yonkers and that was it."

What did they say?

"They made a fuss," he says. "Threatened me."

Weren't you nervous? I ask. With Peg upstairs?

"Nervous, that's a bad thing to be," Jim tells me. "You should never be nervous. Then people will see your weakness."

FRIENDS WHO'D NEVER been to Guinan's sometimes asked if I was ever nervous at the bar. "Well, what about you and Kathryn?" they'd say. "Don't the guys have a problem with that?"

If they did, no one ever let it show when we were around. Even though Peg Guinan was gone, her unwritten rules still held from the days when Buster Coleman could drink with all the white men and feel at ease. The bar was an extension of the Guinans' home, and in that home you did not judge, name-call or make anyone feel unwelcome. Which is probably why no one can ever remember a fistfight breaking out in there. And as long as Jim was sitting in the next room over, his presence felt if not always seen, the same firm Guinan rules still applied.

Not that there weren't a few awkward moments, but ironically they had nothing to do with the guys at the bar.

"Jim thinks Kathryn is your sister," Fitz tells me one day after we've been living in Garrison nearly a year.

No, he doesn't, I tell him. That's absurd.

"Yes, he does," Fitz says. "He just assumes you are. And I'm not going to be the one to set him straight." Heh, *heh.* He laughs at his own pun. I roll my eyes.

I assume the former marshal is just trying to get a rise out of me, but sure enough, a few days later I'm talking with Jim in the kitchen when he asks, "How's your sister?" I stare at him momentarily, trying to figure out what he's referring to, since I'm an only child. "Your sister Kathryn," he says. "How is she?"

I don't know what to say. Eventually I stammer out lamely, Kathryn's fine, feeling totally confused. It's not as if Kathryn's around Jim as much; usually she only comes in on Fridays after taking the train in from the city. Nor does Jim exchange banter and gossip at the bar the way he used to, and well, we are both tall and blond. But surely it's clear she's not my sister.

Later I ask Kathryn, How can he not know? Everyone else knows.

She shrugs. "Maybe he does, and he means *'sister'* "—she makes quotation marks with her fingers. "Maybe it's code or something."

I shake my head. I don't think so, I tell her.

"Well, if you're worried about it, why don't you just tell him the truth?"

Easier said then done, I think. But it bothers me for days. There are no secrets with my own family anymore; we built and crossed that bridge years ago. Jim's not a blood relative, but he feels like family to me.

Then it happens again a few days later while Jim and I are sitting in his living room watching TV.

"How's your sister?" he asks politely.

I swallow hard. Kathryn's not my sister, you know, I tell him, feeling my face get hot. The next thought is habit: *What if he doesn't like me anymore?* I'm a grown adult and this is still hard.

"She's not?" he says. "Oh, I thought she was."

No, I say. I look at the Irishman directly. We're together, but Kathryn's not my sister, I repeat carefully, and wait for him to digest this bit of information.

After a moment of confusion, I see the dawning come across his face. We both turn and watch Walker wrestle some evildoer to the ground. Then a commercial comes on, leaving us without an obvious focus. And Jim says, still staring at the screen, "Well, she's a nice girl."

Yes, I say, silently exhaling. She is.

And that's the end of it.

From that day on, whenever we see each other, Jim now makes a point to ask: "And how's Kathryn?" It's a slight alteration of the question, but it says it all.

IRISH NIGHT.

I arrive late and walk around back where the burning barrel is so I can sneak into the bar without disrupting the musicians setting up outside the front door. The screen door is locked, so I rap lightly until Jane, who is moving at top speed, comes to let me in.

"Hey, hon," she calls, hugging me tight. "Where have you been?"

Working, I tell her. Drove in today and traffic was bad coming out of the city.

"You need a beer," she sings. "Coors Light?" she asks without waiting for the answer. I nod and look around to see who's here. There's Mary Ellen drinking Corona. She waves. And I'm happy to see Clemson, the Ricky Ricardo–looking Spanish-language TV sportscaster who lives nearby and is also a top-notch photographer. I've been bugging him to come shoot some pictures of Irish Night, and he's here with his camera already clicking away.

There's a fine-sized crowd, and Jane is working both the bar and the store alone, wheeling about in a blur of arms, bottles and cash. Mary Ellen and I chat for a moment until I look over her shoulder and see Jane trapped in a deep discussion with some guy while people are clamoring for beer at the bar. I wander over to where she's standing.

Want some help? I ask.

"This gentleman wants a sandwich," she says hesitantly. It's a policy of Jim's not to serve sandwiches during Irish Night so eating doesn't interfere with the listening. But this red-faced guy with Jane is clearly plastered. He's swaying a bit and could probably use some food in his stomach. "I don't know," Jane says. "Should we make him one?"

"Please," he says. He's got slicked-back white hair and a leer. I don't like the leer, but I figure he must be a regular if Jane's even considering his request.

I'll make it, I tell Jane.

"It's your lucky day," she tells the guy and gives me a grateful smile.

"Yeah, it sure is," he says, looking me up and down. I'm conservatively dressed in a short-sleeved black Polo T-shirt, jeans and sandals, but his glance makes me feel like I'm wearing nothing. He hands me his business card—it says his nickname is the Sarge. Great, I think. The Sarge puts his hand on my shoulder possessively: "Hey, I've got some buddies here too. Maybe you could take care of them as well?"

JANE'S GONE, SO it's too late to say no now. I head behind the counter and begin taking out the turkey, ham and bread: two on rye, two on wheat. The Sarge drapes himself across the metal counter while I work, watching me and turning back to his friends every so often and laughing about something—what exactly I can't understand because the musicians have begun a rather raunchy song outside and everyone is singing.

> *Ah, you're drunk,*
> *You're drunk, you silly old fool,*
> *Still you cannot see*

THE SARGE IS probably in his fifties, I figure, and lean with small eyes. Finally he starts getting louder to where I can understand him.

"Yeah, you gotta get 'em while they're still young. Still young and thin." He glances at me and laughs. His friends chuckle, but look uncomfortable when they catch me listening and shift their bodies back toward the drink coolers.

My face reddens. I keep slicing meat. Two rye, two wheat.

What would you like on it? I ask, keeping my voice level.

"Anything you want, baby." The Sarge smiles, a tiny bead of saliva catching in the corner of his mouth.

> *And as I went home on a Tuesday night*
> *As drunk as drunk could be*

Jane peers in: "You okay in there, hon?"

Yeah, I call, still thinking that if the Sarge is a regular he can't be all that bad. Jane goes back behind the bar.

"Look at her. She looks good." The Sarge is now looking straight at me and has moved in closer to where I'm cutting.

I stay focused. You're getting wheat, I tell one of them, because we're out of rye.

"Oooh, a woman who's in control, that's what you want," the Sarge says, not trying to be subtle any longer. "She'll tell you, 'Get on top of me, get under me. . . .'" He reaches in for a high five with his buddies. Clemson meantime passes by snapping pictures as he goes.

"Boy, I wish I were younger. I'd have at her. . . ."

I look over toward Clemson, thinking maybe he'll interrupt before this conversation goes any further. But he just keeps clicking away. Click, click.

> *I saw a pipe upon the chair*
> *Where my old pipe should be*

"Hey, baby, how's that sandwich coming?"

I open my mouth to tell him my name is not "baby," at least not to him. But just before I do, I see Jim out of the corner of my eye, farther down, shaking hands like the mayor, everyone coming up to genuflect and say hello. He's smiling, happy in his environment and this night he created. If I make a scene now with one of his regular customers, it's going to ruin that. I'm here to help, after all.

> *Well, I called me wife and I said to her*
> *Will you kindly tell to me*

"Hey, baby . . ."

> *Who owns that pipe upon the chair*
> *Where my old pipe should be*

I concentrate on transferring the sandwiches to plates.

"How long you been single?"

Look for the pickles.

"I said, how long you been single?"

I shouldn't answer, but I can't help it. I'm not single, I tell him.

"Ohhh, we're too late," he intones.

Yes, you are, I tell him, hoping this will end the conversation. I look behind me on the shelves for napkins.

"*Hey,* baby, *baby,*" the Sarge says, "can I ask you a question?"

I get napkins, put them under the plates and try to total up the bill: $4.25 times four plus 50 cents for cheese. . . .

"Baby, what kind of boys do you like?" He's getting really close, right up against me now.

"Hey, baby"—a fleck of his spit hits my face and I flinch— "What kind of boy are you with?" He grabs my wrist.

I pull my arm away, annoyed. It would be one thing if I were on the other side of the counter as his equal. But since I'm back here serving him, his behavior feels unfair. Like the playing field isn't level.

Hey, I unintentionally echo and it just pops out, I'm not with a boy, okay?

There is silence while they stare at me. Well, I think, that should end the conversation. The Sarge's face squeezes up, confused, as he tries to digest this new piece of information. To put the square peg into the round hole. I hear Clemson chuckle from behind me. More people start to gather around the counter.

I turn back to the sandwiches. But suddenly the Sarge breaks out of his reverie. "Oh-ho, I'm in luck," he crows to his pals. Everyone looks at him curiously, including me. "It's not *boys* she likes." He elbows his friend. "It's *men.*"

One of his buddies, perhaps sensing opportunity, leans across the counter and whispers, "Hey, do you think you could take a break from your job here? I've got some beer down in the boat. We could go drink it together."

Ah, you're drunk,
You're drunk, you silly old fool
Still you cannot see

AFTER FINISHING THEIR sandwiches, I quickly wipe down the counter and go seek out Jane. I'm weirdly unsettled. Not so much because of what the guy said—obnoxious drunks are not exactly a new species—but because he said it while I was ostensibly on the job. He looked at me across the counter, and what he saw really didn't matter. He believed he could say what he wanted, and I'd be reluctant to rebuff him for fear of upsetting Jim, or losing a tip, or even worse, losing my job.

In the back of my mind, of course, I know stuff like this comes with the territory in a profession such as this, like carpal tunnel or boardroom backstabbing in corporate America. Want to keep your job? Learn to deal with it.

But I want to believe that Guinan's is immune to this sort of thing, and that no one who is a regular here would talk like that. And to be honest, my pride is getting the best of me; I hated the feeling of being so inconsequential. So I start chattering away to Jane with a recklessness that reveals just how soft my underbelly is after years in an office cubicle. She listens patiently.

. . . then, I say to her, he's going to "get on top of me, get under me" while I'm making his sandwich for him.

And I wait. Wait for her to say, "What a loser." Or something. But Jane only stares at me with a strange Mona Lisa smile and doesn't say anything. So I'm about to launch back into the story, to explain it again, when the look on her face stops me.

You've gotten this kind of thing a lot, haven't you? I ask.

She nods and squints her eyes. "Yeah, a lot, believe it or not." She's smiling. But her voice is drained.

I look down at the wooden table. Well, I say, trying to redeem myself, at least I didn't lose a regular customer, right?

Jane starts laughing then, for real this time.

What? I say.

She wipes her eyes and shakes her head.

WHAT? I demand.

"Hon, he's not a regular. I've never seen him in my life."

THE SARGE IS standing alone by the metal counter. His friends are gone. So is the sandwich, and it's clearly done some good because he isn't drinking anymore. I walk back into the store, and he looks up at me slowly, trying to focus.

"Hey, baby, that was a great sandwich."

Thanks, I say, and start to wipe off the counter a little too vigorously. I'm annoyed at him but more annoyed at myself. I should say something. He's not a regular after all. Jim wouldn't like this guy's behavior, I think. Neither would have Peg. She would have chased him out of the store with a frying pan. Hell, if Margaret were here, she probably would just handcuff the Sarge to the ice cooler and leave him there for the night.

Nervous, Jim said, *that's a bad thing to be.*

So I drop my rag and put a hand on the Sarge's shoulder and try to channel a little of the detective's AT&T mettle. Look, I tell him, Jane's face flashing through my mind, my name is not "baby" for starters. And I'm sure you're a nice guy, but this "Get on top of me" stuff . . .

I search for the right words.

. . . It's just, well, it's just not the way you should talk to a woman, at least one you don't even know. Especially someone who is waiting on you. This place is a family joint—it's different, people just don't talk like that here.

Then I wait for him to get mad or to start complaining about the insolent help. He stares at me, swallowing a couple of times so his Adam's apple bobs like a fishing lure.

"Well, uh, I'm sorry. I wasn't meaning to offend you."

I let go of his shoulder and start to wipe up the counter again.

"Hey," he says.

Here we go, I think.

"Hey," he repeats. "Well, what's your name?"

I glance at him skeptically.

"What's your name?" he says, then adds, "Miss."

I toss the dirty towel away and face him. I've got no delusions of winning some big battle here—speaking up hasn't cost me anything. In all likelihood the Sarge will forget this by morning. But at the moment, this minor adjustment in his behavior seems like victory enough.

Over the Sarge's back, I see Jane and Mary Ellen in the doorway to the bar making faces and waving at me.

"Miss?"

It's Wendy, I tell him. Then I go join my friends in the bar.

UNDERNEATH THE STATUE of Liberty, the celebrities packed together so densely you didn't even have to look to find them—Madonna, Demi Moore and Sarah Jessica Parker all dining under colored lanterns strung in the trees. Henry Kissinger was there; so was Queen Latifah. A fashionable, boyish-faced man passed us as we stepped off the ferry. That's Tommy Hilfiger, I whispered to Kathryn, our high heels clicking along the sidewalk.

It was August 2, 1999: the economy was still soaring, the dot-com boom not yet bust, and the most pressing news of the day, it seemed, was this new venture called Talk *magazine led by the fabulous editor Tina Brown.*

As the fireworks began, they lit up the twin towers of downtown Manhattan in green and gold bursts, a sight most of us would never see again. Packed together at the gate to freedom, we all watched the brilliant light show dance above the river and world, drawing attention away from the mad currents conspiring beneath.

john's decision

"How COULD you?"

A Garrison full moon doesn't sleep. It climbs into the sky, riding high above the limber tree branches to cast its beams carelessly down onto the earth. On a clear frozen night, a dull glow of white light will spread throughout the valley, creeping around corners and bouncing off ice until nothing is hidden. In the summer, though, the moon exaggerates and taunts. The outline of the granite Osborn castle becomes finer, the great rocks of rolling Nazareth field where we walk Dolly loom larger. And as its pale illumination pours over windowsills and through bare-paned glass doors, the light bewitches the restless into thinking morning is much closer than the hour would tell.

In the late summer of 2002, John Guinan was awake with the moon. And thinking about taking over for his father at the store—not just temporarily, but for the long haul. As the clock ticked toward 3:45 A.M., he tossed and turned beside Mary Jane. His wife had been infinitely patient, accepting that her husband was never in bed when she awoke, never home on a Saturday or Sunday morning. But even she was getting fed up with his uncertainty. "You need to make a decision," she'd told him. "Because if you continue to do this, you need to figure out a way to make it work." It was true, John thought, staring at the ceiling. Something had to be done if Dad was going to stay in the house. And in his gut, John knew what the only option was.

The alarm rang, and with daybreak still far away, he pulled his body from bed, showered and dressed quickly in clothes laid out the night before. Moving through their small rental house, he fed the cats, tossed the rabbit a cracker, then grabbed his knapsack and thermos and stepped into the moonlight.

It should be a simple decision, he thought, his truck's headlights bouncing across the dirt drive down to the main road. After all, he'd always wanted a shot at running the business. Problem is, he would start to feel like this could fall into place, and then it all would creep back up on him—everything that happened with his sister Christine and his father years ago. Then John's blood pressure would start to rise, and he'd wonder if he shouldn't just wash his hands of everything and walk away.

IN LATE FEBRUARY of 1992, while John still worked at the General Motors minivan plant, he and his father went to lunch at an aging wooden tavern called Moog's Farm. At the time, the economic expansion of the mid-1980s had hit the brakes. Corporations were reporting heavy losses and undertaking restructurings to streamline operations and shore up their standing on Wall Street. Thousands were laid off as America shifted from the days of conspicuous consumption to worries about unemployment and debt.

That afternoon at the tavern, John happened to glance up at a TV. At that moment, a local newscaster was announcing that General Motors planned to cut nearly 17,300 workers after posting a record $4.45 billion loss for 1991. A dozen factories would be closed or sold.

The Tarrytown minivan plant where John worked was one of them.

John stopped eating. "Holy shit," he said loudly without thinking.

The bartender came over in a hurry. "What? What's wrong?"

"Jesus Christ, they're closing the plant where I work," John told him.

He stared at the TV while the newscaster laid out the time frame. It wasn't immediate—he had a couple of years while operations wound down, according to GM. John thought about Mary Jane and his kids. He looked across the table at his aging father whose heart hadn't been in the business since their mother's death a few years ago. And a plan started to form.

AFTER LUNCH, JOHN went home and told Mary Jane about what he'd heard.

He'd been thinking, he said to his wife, that maybe he should talk to his father about taking over the store. It was a dream he'd long had, lying awake at night, mulling what changes he'd make, improvements and that sort of thing. But the timing had never been quite right.

Until now, he said.

Mary Jane listened. She'd known the Guinan family most of her life and had watched how difficult it had been for them to all get away and do things together. The store, she knew, was a full-time commitment—twenty-four hours a day. Most people didn't realize that. They just saw everything in action for the few minutes they breezed in and out. At the post office, she knew that when her workday was done, she could leave it all behind and come home. At Guinan's there was no distinction between work and home. Mary Jane loved the store, she respected what Jim and Peg had built, but she did not want to carry on Peg's role there. She and John had their own life together, different from his parents' perhaps, but it was working.

Still, in the back of her head, Mary Jane had always figured this day might come with her husband. It was in his blood, after all, and no matter how much he tried to distinguish himself from his father, the older John got, the more similarities she saw. Watching him now talk so passionately about the store and bar, she knew that if he wanted this, she would get behind him.

OVER THE COURSE of the next year, John and Mary Jane took his father to dinner a couple of times. John told his father that when he was ready to retire, he should let John know. As is John's live-and-let-live way, he left things very informal, never pushed his dad too hard, never laid down any firm calendar or transfer plan. He wasn't in a rush, after all.

Meantime, Mary Jane says, while she made it clear to Jim that she wouldn't be coming to work at the store herself, she nevertheless fully supported John. It's just that things would be different from how they were when he and Peg ran the show. She knew this probably would not go over well with her father-in-law. But as Mary Jane and John remember it, Jim would just nod in a nonconfrontational way and that was it.

Right about this time is when Jim and Cliff were beginning to make their afternoon lunch excursions to Jeremiah's. It had been four years since Peg died; Margaret had moved out, and Jim was spending less and less time at the store and more and more time in the staff car tooling between his favorite bars. He was tired and ready to give up more of the day-to-day responsibility for the store and bar. Yet the notion of the store continuing without a mom as well as a pop overseeing things didn't sit well with him. It just wasn't the way things had been done, you see. "Peg and I, we were a *team*," Jim will say, head held high, even to this day.

And so, in the fog of all the change, all the "set 'em up" lunches and all the little ways the still-fresh loss of Peg touched him alone, Jim took Mary Jane's own reservations about the store to mean she didn't really want John there either. And he made a decision of his own.

IN THE SUMMER of 1993, Christine went to a doctor's appointment and left her sons with their grandfather. When she returned to pick one of them up for a basketball game, Jim suddenly wanted to sit and chat.

They took seats at the kitchen table. And as Christine remembers it, her father said, "I've been thinking about this. I'd like to go visit some relatives and take a vacation in Ireland. Would you and Mike be interested in taking over the business?"

Christine's husband, Mike, had his job then in sales, but the idea of running his own show at Guinan's appealed to him. Christine, the only Guinan child to have finished college, had a good job working as a computer consultant for IBM. However, the very same week her father approached her, she says, IBM offered a voluntary buyout plan. "I thought my mom was trying to tell me something," Christine says. "That this is what Mike and I are supposed to do. So I went back to Daddy, and he said he had talked to everyone. Margaret and Jimmy obviously didn't want any part of running the store. And that it was the same old thing with John. John wanted the business and Mary Jane said no."

ALL HELL BROKE loose not long after when John was on his way home from the Tarrytown plant. He walked into the bar to grab a cold beer and found his father sitting inside with one other guy. As he pulled a bottle from the red cooler, John mentioned to his father offhandedly some complaints he'd been hearing about the attitude of a part-time employee at the store. And his dad told him, "Well, that'll change in another month."

John looked at him, confused, and said, "What are you talking about? What's going to change?"

And that's when, John says, Jim told him about Christine.

"SHE'S WHAT?" John yelled. "WHAT ARE YOU TALKING ABOUT? Mary Jane and I took you out to dinner all these times and this was a done deal, it was set. What about the Irish tradition of passing the business to the first son? I CAN'T FUCKING BELIEVE THIS."

He screamed for a while longer, and then slammed his beer on the floor and left.

THAT NIGHT JOHN called Christine on the phone.

"When were you going to TELL me," he demanded.

"John, I thought you knew," she told him. "I would never have done that. I'm your sister."

John wasn't listening. "I have no sister," he screamed, slamming the phone down.

Christine called back, crying, but John wouldn't pick up. She called Margaret, who said she knew nothing about any of this. Christine kept calling John back until finally Mary Jane picked up. But John still wouldn't come to the phone even though Christine could hear him yelling in the background.

"Tell him to go ask Daddy," Christine told Mary Jane. "Ask him what happened."

The next day when John went to work, he hadn't slept at all. His eyes were red and the guys at the plant tiptoed around him, warning one another, "Man, stay away from John today."

Finally John called his father on the pay phone during his break and started yelling all over again—"How COULD you? We took you out to dinner so many times, Mary Jane and I."

"I didn't mean to do it like this," John recalls his father saying. "And I told him, 'You made your bed. Now you lie in it.' "

John hung up because there was a long line of guys waiting to use the pay phone. His heart was still pounding, so he went over to a pile of new lumber, picked up a hammer and just started swinging away at it until his manager came and asked him to calm down.

Right then, John thought he never would.

Jim tried to fix things in his own way. He called a meeting and told his children, "You should be able to work this out." By year's end, Christine and Mike were slowly taking charge and tentatively asked John if he wanted to put any money into the store. Or did he want to come work a few hours during the week? John tried that, coming in on Sundays for a while. But his heart wasn't in it, and besides, it was hard working for his sister because he felt like it should have been his all along.

It wasn't the Guinans' style to sit down and rehash everything. Margaret stayed out of it; so did Jimmy. Mary Jane meantime waited, hurt, for everyone to realize that her feelings had been misinterpreted. But that never happened, and so she and John drifted more to the periphery of the family and the store. "We successfully avoided each other" is how Christine puts it. With no outlet to make sense of what happened, John's anger just froze in time.

The years passed. John finished his time at the GM plant, and Christine and Mike ran Guinan's profitably until 1997 when their own battles with Jim finally drove them to leave. Soon after, they moved to Florida to care for Mike's sick uncle. After they left, John showed no desire to return to the store and pick up where his sister had left off. Instead, he got a job in landscaping that allowed him to be outside as much as he wanted. His boss, Lew, was older, a strong, wiry, curmudgeonly type who took John under his wing and respected his work ethic and skills. Most days they were buddies, cursing and laughing and drinking a beer outside after work. Sometimes, though, when Lew was encouraging John as he scaled a tree, chain saw wrapped to his waist, the relationship was more like father and son. The incident with Christine did fade, and John might eventually have moved on altogether, except that his real father got sick. And so he and his sister Margaret had abruptly ended up behind the counter filling in. Now all the old anger and hurt had returned, still preserved in the moment years ago when he walked into the bar and heard the news about Christine.

An apology from his dad was probably out of the question. After so much time—the misunderstanding would stand. Waiting for thanks for his hundreds of mornings of lost sleep seemed unlikely too. John was now behind the counter because his father needed him to be there. And for a proud older Irishman used to doing the asking and the telling, to *need* his son's help was a tough reality to swallow.

So John now had a decision to make. He had to decide whether he could take the torch after all that had transpired. To do so meant finding a way to forgive his father, not so much for the things he had done and said, but for those that he hadn't— and probably never would.

"*Don't forget to call your mom back,*" *Kathryn said, stepping over the suitcases strewn across the floor.* "*She called last night. Wants to know how the trip was.*"

I know, I know, I said absently, flipping through the newspaper, trying to figure out the day. It was a clear morning, warm for September 11. Joggers were running by outside our apartment window. I should be exercising too, I thought. From our tenth-floor window, I watched the workers stream off the ferry, walking to their offices in the World Trade Center towers and other downtown office buildings. Maybe I'll get to the gym at lunch, I thought. Swallowing a final sip of coffee, I headed for the shower.

"Okay, but you always forget to call her back, and then she thinks I didn't give you the message."

Kathryn's right, I thought, I should just call now. Then I checked the clock: 8:40 A.M. No, too much to do. I needed to get moving.

Look, I'll call her later this morning, I shouted, turning on the water.

back to ground zero

See, there's this place.

Walter drives me to Sears to help pick out a new vacuum cleaner. On the way, he lectures me about the mechanics of wise spending. "The way to know if you should buy something," he says, puffing on his cigarette, "is to ask yourself three questions: Do I need it? Do I really need it? Can I live without it?"

I think about this as we chug along in his 1990 rusted-out blue van with lumpy seats, which he insists on keeping despite the fact that he also owns a perfectly comfortable Saab. Then I say to him, more out of curiosity than anything—

But what if you just want it? What if it makes you happy?

I feel him accelerate a bit. "Happy?" he says. "HAPPY?" The van starts to shake with the speed and he slows back down.

Yeah, I say. Happy.

He chews on this momentarily and then shakes his head. "Well, I don't know. I never thought about that."

We pull into the Sears parking lot, where Walter circles, looking for the most opportune spot. After we park, I unstrap my seat belt and look over at him. He seems nonplussed, muttering under his breath, *"Happy?"*

Don't worry, I tell my new friend. Unless you're coming over to clean, I really *need* a new vacuum.

OSBORN'S CASTLE ROCK taunts those below. During heavy cloud cover the entire estate can disappear into the hillside and then reemerge as if by magic when the sun wakes up and mist burns away. No one's actually sure who, if anyone, still lives there, although people at the bar talk about the crazy caretaker who stays there day and night with wild dogs, chasing away trespassers with whatever tool he happens to be brandishing at the time—be it maul, ax or shovel.

Refusing to believe the talk, one afternoon I take a chance and drive halfway up the unmarked road leading to the estate. Ever since Frank the preacher told me the lore about the earth's magnetic pull up there, I've wanted to see for myself. Unfortunately, I've chosen a foggy day to make my trek, and I can't see well around the turns. About midway, I stop to get my bearings and from the distance hear barking. Maybe I'm imaging things, but the fog seems to get thicker and thicker and the barking closer and closer. I hesitate for only a moment more before putting the car into reverse and hightailing it off the mountain, half expecting, as I go, to see a man with a pickax looming behind in my rearview mirror.

SUMMERTIME IS PEAK tourist season in Philipstown, a chance for merchants to make up for the sleepy winter months when they pore over the books, adding and subtracting and furrowing their brows, wondering how they'll make it to spring. Visitors descend with picnics upon Garrison's Boscobel, a restored Federal period mansion flanked by herb and flower gardens and postcard-worthy views. Cold Spring's quaint main street, with its rows of antique stores and sweet restaurants, thrives as crowds surge up and down the sidewalks on weekends marveling at the lush green cliffs rising from the river. With winter's strict gloom far out of reach, they pause at real-estate agents' windows, pointing at ads and imagining buying a little piece of this perfection to keep.

These days, more and more, I find myself pausing with them. Our rental home's yearlong contract will be up soon. It's been nearly a year since the attacks, and returning to the city is on the horizon. Not long ago, I'd received an e-mail from our landlords wanting to know our intentions for renewing another year.

"Well, we can't, obviously," Kathryn says when I read it to her. "Renew, that is." I must look surprised because she says softly, "This was supposed to be temporary, remember?"

She's right, of course. Garrison was initially just an unexpected layover on the way back to town—a place to gather our wits and belongings before plunging back into our old lives. The commute is more than Kathryn, who makes the trek most, bargained for that first afternoon we sat in Guinan's, hypnotized by Jim's tales. Friends, meantime, seem increasingly puzzled that we've stayed here this long, and I can't exactly blame them. A year ago, I couldn't have imagined wanting more than what New York City had to offer.

But a year ago I never knew the quiet lure of what happened out here—of Jim predicting the first snow, the burning barrel and Lou-Lou's mouth wrapped around her Monday morning roll. Driving past the dramatic slope of Sugar Loaf Mountain, I would think about leaving all this behind and unexpectedly feel my throat catch.

Which is why I pause at these agents' windows to stare at the addresses on dirt roads whose twists and turns are now as familiar to me as those around Wall Street. I think about my small savings from a decade of investing in nothing more permanent than a money market fund. I imagine making a commitment to this world not just temporarily, but for longer; I even think about how good the schools are. And then I do battle with that old internal clock that tells me distractions such as leaky pipes, diapers and a yard to mow are not yet on the agenda.

All this, it reminds me, *tick, tick,* is supposed to come later.

So I stand peering through glass alongside the tourists, paralyzed, and feeling like a jealous lover as I silently tell this town I want to be with her unconditionally—winter, spring, summer or fall.

ONE EVENING JANE and I are behind the bar. Fitz is there. So are Kathryn, the Preusser brothers and the one I call the Count. It's a little after 7:00 P.M. when a West Point cadet comes in. He's clean-cut and proud, back ramrod straight—a first-classman poised to graduate. He's wearing his heavy class ring, which he glances at every so often as if to make sure it's still there. He sees us all hunched together in our familiar pack and goes to stand by the window alone.

Kathryn whispers to me, "I feel sorry for him standing there alone."

So I call to him—Come over here. Sit. We won't bite.

He walks over, smiles gratefully—"Okay"—and edges into our fold.

Fitz sizes him up. "Are you in the military?"

"Yes, sir," the cadet answers respectfully, as if he can tell Fitz is a fellow soldier.

Fitz protests—"Oh, don't call me sir." But I see the veteran sitting up just a little straighter himself now.

"Were you in the military, sir?" the cadet continues.

"I was in the army," Fitz tells the boy, smiling at his insistent "sir." Fitz continues: "The Rangers. Before you were born." He pulls a card out of his wallet: William Fitzgerald, Number 1332. The two talk for a while, wrapped up in their own world. The cadet is flying to Fort Bragg, North Carolina, not far from my hometown.

Jane hands the cadet a Rolling Rock, which he sips slowly. Far too slowly for the Preusser brothers. "Come on," John says. "Let's go. Train's here in three minutes. We're clocking you. You've got a big night in the city. Let's get started now."

So the cadet swigs. Long, controlled swigs. We all watch.

"So where are you staying in the city?" John asks after a moment.

The cadet swigs again, same long draw. Wipes his cherry-red lips. The beer is half gone.

"I don't have a place."

"Ooohhhhh." John grins his conspiratorial playboy smile. "Just going to make your way, huh?"

"No," the cadet says firmly. "I'm engaged."

"To someone you met in high school?" Kathryn asks.

"No," he says again. "Someone I met in church."

Right, right, we all murmur, quieted by the earnest declaration. He keeps swigging the beer, shirt tucked in, manners impeccable. He turns back to Fitz, as if he's decided this is the man most worth speaking to here.

"How long were you in Vietnam?" he asks Fitz.

"Not long," Fitz says stilly.

The cadet looks at him curiously, but has the good manners not to ask the obvious follow-up question. Still, it's apparent he wants to know why Fitz's tour of duty was so short. So Fitz, after taking a long swallow of his beer, breaks his traditional silence on this matter and decides to give the kid a rare glimpse beyond the snapshot.

"I was shot."

The Preussers look up, but don't say anything. Neither do Jane or Kathryn or I. We all know this information, but are surprised to hear Fitz talk about it so openly with this kid, a stranger. Still, it's obvious there's some connection between them, even after only twenty minutes.

"I was in about ten months," Fitz continues matter-of-factly. "I got shot three times. . . . One bullet is still there."

In your knee, I mumble, mostly to myself, walking out from behind the bar.

The cadet is listening intently, with a strange look. It's not fear, though. In fact, he almost looks proud. And he clearly wants to hear more. I see Fitz study the kid's face closely, making sure he's not upsetting him. The cadet doesn't break his gaze. So Fitz tells him more.

"The bullet went in there," he says, speaking directly to the cadet now, the rest of us sort of fading away. Fitz rolls up his pants, showing the back of his leg where a perfect scar runs around the bend of his knee and out the front.

"Where were the others, sir?" the cadet asks.

"One went in my hip. Took off a chunk of it. But I didn't feel that. Then another came in. Hit me right here." He points to his groin. "And it came out through here," he says, pointing to his backside.

"I thought it was going to tear me a new asshole," Fitz tells the cadet, with a tough smile. Then he takes a sip of his beer and adds, as an aside, "But it didn't."

The cadet smiles back. This is privileged information, and he knows it.

And just like that, by some mutual agreement, they are done talking. The cadet finishes his Rolling Rock while we all resume our normal patter. John buys him another one for the train. As he makes ready to leave the bar, the cadet stops directly in front of Fitz.

"Sir," he says, nodding respectfully. Fitz nods back.

"And when were you in Vietnam again, sir?"

Fitz reminds him.

"Thank you, sir," the cadet says. And he's gone.

We are all silent for a minute. The cadet is about to enter a part of Fitz's world that none of us will ever truly understand no matter how many beers and afternoons we share down here.

"A year from now," Fitz says, breaking our silence, "he'll be sucking sand somewhere." He says it flatly, as if to shake off any trace of emotion we might pick up in him.

I look away. Did anyone get his name? I ask.

"Yeah," John says. "Cadet-dick."

We laugh. Even Fitz.

I'm thinking out loud when I say, He could be dead in a year.

Fitz turns full body on me. "Don't say that," he commands. I must look surprised because he continues, "Knock on wood, right now." He points to the bar.

I knock on my head instead, hoping he'll think that's funny.

"No," he says fiercely, grabbing my hand. "Knock on wood. Right now."

There's no kidding around anymore. I knock on the bar.

"Don't ever say that," he says.

But we don't even know his name, I say nonsensically, a little frightened by Fitz's reaction.

For a moment Fitz holds my hand to the bar, then suddenly lets go, almost like he's deflating. "He was so naive," Fitz says, his anger gone just as quickly as it came.

"A baby," adds Ed Preusser.

"But what would you do without him?" Fitz asks rhetorically, rolling his beer bottle around on its coaster. "You need him. You couldn't do your job without him. How could you go to work if he wasn't there?"

We all consider that for a moment. Images of the fresh-faced soldiers who led us back to our battered apartment building after September 11 flash through my head. If they were scared that day, scared of what was to befall them in the months to come, they never showed it. They simply did their job, which was to protect and assist us, to help us get our cat and begin resuming our lives as normally as possible. I didn't know their names either.

Just then Mary Ellen comes in. Jane has a Harp open before she reaches the bar.

"I don't want one," Mary Ellen protests, though not too hard.

"Too late," John tells her. "It's already unbuckled."

I sit down tentatively next to Fitz, wanting to make it better, wanting to say I'm sorry and that I wish I understood more. But I know I can't do that in this place. Like Fitz always says, it gets worse if you rehash it. The veteran looks at me hard for a moment and then puts his hand on my shoulder protectively.

"And get one for the Gwendol too," he tells Jane. "In fact, this round's on me."

WALTER IS SCHEDULED for a physical—his first real one in thirty years. He hasn't been eating well for months out of nervousness and has dropped about twenty pounds. "I don't *want* to know if anything is wrong," he says. A decade ago, Walter had Lyme disease from a tick bite; it started with his knee swelling and worsened until he couldn't walk any longer and had to quit his job in sales. For two years, the only time he left the house was for physical therapy and intravenous feeds of medicine. Every daylight hour, to prevent his muscles from atrophying, Jos would walk him around the house, his large frame leaning on her shoulders. Walter never went back to work full-time—thanks to his lifetime of frugality, they had enough to live on—but the whole ordeal left him petrified of illness and doctors.

I pull into his driveway just as he's about to leave. I'm going with you, I tell him.

"Oh," he says, thinking about this. "I thought you had to take your car to get fixed."

I shake my head. It can wait, I tell him. Want me to follow you? I ask.

"Well, it's silly to take two cars," he says, shifting his weight back and forth. "Waste of gas and all."

Get in, I say.

Throughout the drive, Walter sits in the passenger's seat rubbing his thumbs together nervously. He's brought three pairs of reading glasses and his cleaning solution to keep him occupied in the waiting room. We arrive early and go to a Starbucks next door, where Walter insists on paying for my three-dollar latte, which tells me he's truly a wreck.

"Look," he says when I appear surprised, "this coffee's going to make you *happy,* right?"

He smokes one last cigarette and then we go into the doctor's building. Walter wants to take the stairs, but I pull him into the elevator. We're only going to the second floor, I tell him, adding—If it crashes, we won't even feel it. He stares straight at the ceiling until the bell dings.

Inside the waiting room, the receptionist gives him two clipboards to fill out. "One for me and one for you," he says, handing one over. We go over each ailment on the sheet: "heart trouble . . . no, diabetes . . . no, depression . . . I guess so . . . stomach pain YES," he says loudly.

There's a space that asks for an emergency contact. He pauses. "I guess I'll list you," he says. "Right?"

I nod.

Finished, we sit together waiting. I flip through an issue of *New York* magazine while he studies *Better Homes & Gardens Wood* magazine. "This is what you should be reading," he says, tapping a finger on an article about termite prevention. After a moment, he gets up to examine the doctor's diploma, probably to see if it's genuine.

"Mr. Johnson?" A nurse appears from behind the frosted glass door.

"Oh, okay, that's me," he says, though it's clear he'd rather that not be the case.

Walter, I say as he's about to disappear behind that door. He looks back.

I'm right out here, I tell him.

"IT WAS GOLDEN."

We're driving home and Walter is crowing about the quality of his urine sample. "That's what the doctor said." My neighbor has checked out perfectly fine, and the doctor has even given him a new prescription to calm his stomach a bit. "I told him about all the candy I've been eating, and he couldn't believe it. But can you believe what that guy charges? I had to pay that right out of my pocket. . . ."

I smile and head back toward Garrison.

DAN THE ATTORNEY stops me as I'm leaving the store one morning.

"Can I talk to you for a minute?" he says.

We walk a few paces together; he seems uncomfortable.

"Someone approached me on behalf of one of the tenants' association members about Jim," he says finally.

We stop walking. I'm facing back toward the store. Well, what does that person want? I ask.

He rubs his forehead. "They wanted to know if Jim would be amenable to accepting a stipend to live somewhere else, if he would close the store and relinquish control of the building. It would be sort of a buyout. This person wanted me to act as an emissary, even though I don't think this request was sanctioned by the overall board."

Behind him, Jim and Margaret are leaving the house for a doctor's appointment. Jim limps slowly to Margaret's black convertible, his white socks blaring from his sandals and khaki pants. He sticks a few letters in his green mailbox. Then Margaret helps him ease slowly into the passenger's seat. I watch this happen over the attorney's shoulder and feel a surge of anger toward the one or two people who keep these rumors alive.

Do you think Jim would do it? I ask finally. Leave?

"I went to Margaret," the attorney says. "She said, 'This is my father's home. It's where he wants to die.' "

IT GETS HARDER to leave myself—Garrison, Jim, John and mostly Guinan's. Every time I do, I'm afraid it may all disappear before I get back, which is crazy, of course, but everything feels so fragile. If I'd been something else, a lawyer, investor or politician, I might have had a useful skill to offer the family. Not that anyone at Guinan's was asking for help, but that was how this place made you feel: like you wanted to give back as much as it gave to you.

Kathryn comes up with an idea. We talk it through one late afternoon, sifting through the pros and cons while sitting outside watching Dolly roll in the ivy. Before the sun sets, we head inside the house, and I send an e-mail requesting a meeting with the head of our newspaper.

Later that week we drive into the city. It's late summer of 2002, and the *Wall Street Journal* staff is moving in shifts back to our downtown headquarters next to Ground Zero. Kathryn is scheduled to go today. We pass Chambers Street. On the right is our old movie theater, posters of missing September 11 victims still plastered across the metal chain-link fence. To the left is the pit. There is our fruit man; he looks the same. So does the coffee cart vendor. The pauses before each street turn are familiar. There's our pharmacy, dry cleaners and nail salon. My foot on the clutch and gas pedals, instinctively maneuvering in old familiar ways. Our old apartment building, now reopened, straight ahead.

Clutch, gas—and feeling like I'm stepping onto a stopped escalator. You know it's stopped. But when you step on the stairs, you still stumble, expecting the motion. Stumbling a little now. Feeling like my entire life is still contained in these few blocks. Knowing that it's not.

THE FIRST BAR turns me down for a drink. It's 10:30 A.M. and the guy behind the counter at the Ear Inn, the closest thing I know around here to a Guinan's, has his doubts—even about a clean-cut blonde in khakis. So I wander through the streets, subway heat seeping through the grates onto my hot sandaled feet, until I find another open spot. Sosa Borella, 460 Greenwich Street. I don't know the place; it looks a bit too upscale for what I need. People are getting coffee and bagels. But there is a bar.

I don't see any beer readily available. I'll have a scotch, I tell the guy. He doesn't know I don't drink scotch—or even like it. Glenlivet, neat, I tell him, calling out a phrase I've heard my friend Erle use.

I smile apologetically as he hunts around for the bottle. Kind of early for this, I know, I say.

"Oh no," he assures me too quickly. "If I could have one, I would."

He thinks I've got a drinking problem. I consider this, and then take a sip anyway. It burns my throat, but my stomach relaxes. I take another sip and set aside the rest—it's Guinan's I really crave, after all, not a drink. I've only got a few hours before my meeting. And I need to think.

"WHAT CAN I do for you?" our paper's managing editor asks.

I clear my throat, shifting in the chair. I'd first met Paul Steiger when I was twenty-two years old writing about Eastman Kodak and had flown to New York City to cover my first shareholders' annual meeting. It had been the beginning of my career and early in his reign at the *Journal*'s helm. At the time, we sat in his office and talked about my future with the paper, all the possibilities that lay ahead.

Now, nearly a decade later, I feel like one of the commuter trains about to derail.

He's waiting.

I want to leave the paper for a while, I tell him, hoping he can't smell the two sips of scotch on my breath from earlier in the morning. He watches me, eyebrows cocked, so I continue. See, there's this place, I say. It's a tiny hole-in-the-wall spot up north and . . .

It all tumbles out. About the bar, and Jim and John and Margaret trying to hold it all together. And how I'd found myself behind the counter, meeting old Ken Anderson and about Fitz and how I knew it wasn't exactly what I'd been doing at the paper, but did I mention Lou-Lou? . . .

So maybe there's something there, I tell him. Something to write. A story. I don't know. But I think, I think I just want to be there for a while. . . .

I trail off and look down, waiting for him to speak. Waiting for the polite "I'm sure it's a great place, but maybe this isn't the best time? There are lots of opportunities opening up here. . . ."

He tucks his hands behind his head and leans back in his chair. "Go," he says.

My stomach turns over. He's pissed.

"You should go," he repeats more gently. "Go be there. I understand completely."

WE REALLY DON'T have time to do this, I told our friend Jessie, as she lit another cigarette. It was November 1, almost two months after the terrorist attacks, and Kathryn and I were standing outside some ramshackle country store in the middle of nowhere. There was no sign, just a green mailbox that said GUINAN.

"Trust me," Jessie said. "You guys have gotta see this place."

We can't stay long, I told her.

"Just one beer," she promised.

part

III

But now the storm is over

And we are safe and well

We'll go to a public house

And we will drink our fill

We will drink strong ales and porters

And make the rafters roar

And when our money is all spent

We'll go to sea once more

—"The Holy Ground"
(Irish seafaring song)

human duct tape

You've got to live for today. . . .

L ate 2002.

The cold's coming and everyone is battening down for the winter. When the wind blows too hard, John and his partner, Lew, can't work outside, so John begins spending more time at the store. He and his father edge around each other staking out their individual turf. When Jim hobbles out into the store to get some bread from the cooler for his toast, John yells out, "Jesus Christ, would you just sit down. If you don't stay off that foot it's never going to get better."

There's a store full of people who look up nervously at Jim, who stands there while his son reprimands him. They glare at each other. "I want some toast," Jim says finally.

"Well then, ask for help, would you?" John hollers back.

Jim pauses and then with a mischievous smile cries out, "HELP!" We all chuckle.

"Don't encourage him." John scowls at us all and goes back to buttering rolls.

EVEN FROM THE outside, the house for sale has clearly seen better days—the brown siding is stained black in places from neglect, its deck is rotted through and the bluestone front porch is cracking off in chunks. There's a garbage bag taped where a window should be. And even the chimney top looks pretty shaky from where I'm standing with Melissa, a dry-witted real estate agent we'd met during our first weeks up here. Kathryn and I have gone on a month-to-month lease with our rental and are starting to look at want ads for the city. With time running short, I've moved from staring at Philipstown real estate flyers to actually staring at houses themselves.

Despite its many issues, this one of Melissa's has some potential. It sits down in a small valley with a fenced-in yard—which Dolly would like. And there's a rambling stone wall that looks really old.

Is that from the Revolutionary War? I ask Melissa hopefully.

She rolls her eyes.

Still, it's hard for me to imagine learning to take care of all this. A manageable apartment in New York City is one thing, but this struggling structure . . . it's like a needy baby. And while Kathryn's not standing in the way, she's made it clear that because this is my dream, I'll be the one primarily responsible for its care and feeding. Right now that seems rather dangerous, since Melissa is talking about buried oil tanks and well-water psi, and I'm not understanding a word of it.

". . . And I'm not sure the last time the septic was emptied," Melissa tells me.

The septic?

THE WEATHER REDUCES everyone to platitudes, or at least their own versions of them.

"It's colder than a well-digger's ass out there," pronounces

John one morning, stamping his feet after hauling up beer from the cellar.

I don't get that, I say, holding the door open for him. I mean, you're always using that expression, but what exactly does it mean?

"Hey, Dad," John calls to his father, who is eating a soft-boiled egg and reading his local *Journal-News*. "Tell Wendy what it means when it's colder than a well-digger's ass."

Jim peers at me over his teacup. "That's cold," he says matter-of-factly.

Snow in the city is beautiful for the first few hours. Then traffic churns it to thick black slush while the merchants and snowplows engage in warfare, one shoveling snow onto the street while the other shoves it back toward the sidewalk. Out here in Garrison, everything stays white and quiet for weeks on end. When a new snowfall comes, it softly touches up the landscape's melting canvas, nature restoring her original design. One morning, I'm walking into Guinan's and turn my face up toward the cool sting of flakes. Bruce, a good-natured packaging executive, passes me with his morning papers and says hello.

It's really coming down now, I tell him.

"I've never seen it go up," he says.

JIM IS POKING around the coffee station in his burgundy cardigan sweater, cutting handfuls of straws in half to make new coffee stirrers since the store has run out. "It saves money," he tells John when his son asks him what the hell he's doing over there. After a moment, Jim walks into the kitchen and returns with a card, which he hands me. I open it and inside beneath the message *May all your Christmas dreams come true,* it says: "To Wendy, Thank you for all your help. Merry Christmas, Jim."

There is fifty dollars cash inside.

"It's your bonus," he says, patting me on the shoulder. "Merry Christmas."

I swallow hard and try to hand him back the money—You don't need to do this, I say.

"No," he says firmly. "You've earned it."

That night I go home and put the card up on my office wall. And then I take the fifty dollars cash and tuck it inside a special compartment in my wallet. My own emergency money, I figure— like my grandfather's glove compartment change, only adjusted for inflation.

Later I get an e-mail notice from work telling me that because of the lousy advertising environment, bonuses are only going to select employees this year. I am not one of them. I read the message a few times and then hit delete. It's thousands of dollars, but I can't argue. The truth is my heart has been elsewhere this year.

I'M SITTING IN the bar with Margaret; the detective is drinking coffee and I'm sipping a beer.

The detective is doing her best to make polite conversation, but she keeps staring off at the shifting ice along the banks of the river. "You look tired," I tell her.

"I just can't do this anymore," she says, her lips forming a tight line. She doesn't look at me. "I told John I'm done here after the holidays. I want my life back," she tells me, voice cracking.

You can't quit now, I want to say, *what about all of us?* But I say nothing. Because looking at her exhausted face, I realize I have no right to want this place to stay here as much as I do. The burden is not on my shoulders.

I hand Margaret a five-dollar bill, but instead of reaching into the old bar box for change, she walks into the store, motioning for me to follow. I wonder if she's going to tell me that I can't drive.

I'm fine, I start to say.

"Here," she says, interrupting, and presses my five-dollar bill back into my hand. "This one's on me, and besides"—she smiles—"you know I can't sell beer as a police officer."

I pocket the money awkwardly.

"Look, I've got to go get dressed for work," she tells me. "Can you watch the register for a minute?"

It's the first time Margaret's ever asked anything of me. I nod, and then remember it's Wednesday when the store and bar shut early at 4 P.M., a leftover rule from the days when Jim took the afternoon off to golf with his Fearsome Foursome. Look, I tell Margaret, why don't I stay for another hour until the store closes—keep your dad off his feet?

She hesitates for a moment, weighing the offer. Finally she says okay and goes upstairs. When she comes back down, she's wearing makeup and a long coat with dress pants and a flowered shirt. She looks pretty but drained.

After she leaves, Old Mike comes in early for his Schaefer. It's three-thirty and he's spent most of the day outside in single-digit weather while his crew fixed telephone lines at West Point. He reaches gratefully for the beer to chase away the cold. Colonel Tom drops by too. Hearing him, Jim comes in from the living room and pulls up a stool. He buys us all a round and asks for a nonalcoholic Haake-Beck for himself. Occasionally a customer stops into the store, and I go ring them up. But mostly the four of us sit together swapping stories and laughing as the afternoon ticks toward 4 P.M.

MARGARET'S DECISION IS the final straw. John tells his family that he plans to take over the business. Then he swallows his pride and tells the Station Plaza board of directors the same thing. Asks them to put his name on the lease too. The directors don't agree immediately; instead, they ask him to make a pro-posal about the things he wants to do with the place and submit it in time for the next board meeting.

John comes to my house, and I start typing up his ideas on my laptop. It's hard to get him to focus. He keeps staring outside, pointing out different evergreens, telling me stories about his tree-climbing expeditions.

Finally I shut the laptop's lid. How do you feel about all this? I ask him.

He shrugs. "Hey, kid, you know the rule. You do what you've got to do. Now, see how those tall trees have most of their branches on top? They're called javelins, and they grow that way to get the most sunlight. . . ."

WHEN PEOPLE IN town hear about John's plan to take over, they offer to help however they can so he can keep up his land-scaping job with Lew, which he can't afford to give up yet. Col-onel Tom says he'll take a bartending shift on weekends. John, the mortgage broker up the road, offers to help with a business plan. So does Clemson, the Latin TV sportscaster and photog-rapher. Dan the attorney promises legal advice. And bit by bit the human duct tape that keeps this place together tightens its hold.

One morning a scruffy guy with blond whiskers and bright blue eyes shows up in a blue wool cap, black sweatpants and work boots. When he takes off the cap, it looks as if he hasn't combed his hair in weeks. He pours himself a cup of coffee and sits at the bar with a copy of the *Wall Street Journal*. He doesn't bother to introduce himself to me.

I'm Wendy, I say finally.

"Robert," he says in a monotone voice. "Come to help out John some in the mornings."

I size him up, figuring him for a down-on-his-luck barfly look-

ing for part-time work. I don't think John can pay, I tell him protectively.

"Oh, I don't want to get paid," he says, looking over my shoulder. "Good morning, Jim," he calls in the same low-key voice.

"Well, well. Good morning, Robert," Jim says, peering
around the stove into the bar. "Haven't seen you around here in a
while. How've ya been?"

"Not bad," Robert says. "Just enjoying retirement. Playing
tennis. Taking the dog for a walk."

I look back and forth between the two of them, surprised by
the familiarity.

"That's grand," Jim tells him, picking up the boiling teapot.
"Just as you ought to be doing."

What are you retired from? I ask Robert skeptically, trying to
imagine this guy ever gainfully employed.

He scratches his matted hair. "I was a French and Spanish
teacher down in the Lakeland school district. Been coming to this
place since the early seventies. I was pretty poor then; didn't take
care of myself. Peg used to feed me. Jim taught me to play darts.
There was always someone here who knew something about
everything—lawn mowers, how to run pipes underground. Anyway, I heard through the grapevine that John might need a hand, so
I figured I'd come down for a few hours in the morning. Not doing
much else since I retired. Well, like I said, playing tennis and walking the dog. But that's about it."

He takes a sip of his coffee and looks at me mildly. I blush,
having so obviously sold him short based on his scarecrow appearance. Fortunately, he doesn't look the least bit bothered,
and we get to talking.

Robert had grown up in New York City's Hell's Kitchen, a
rough domain in the 1950s and '60s. He, his parents, his
grandmother and five siblings all packed into a boxy four-room
apartment with low ceilings and one bathroom. They slept

three girls to one bed, three boys to another on two twin mattresses, with the littlest brother getting the crack between them. Rent was $52 a month. Robert's last name, Witty, is made up—something his father picked out of a phone book because it was easier to spell than the family's real Polish-Russian name. "I don't even think I could spell it for you now," Robert tells me.

As the oldest, Robert started working young, selling nickel newspapers on Broadway, doing anything he could to get out of the cramped living space where his siblings were roller-skating, bouncing balls, pummeling each other with brooms. He worked as a ticket taker in the theater district because his mother was an usherette and could sneak him into shows. In tenth grade, Robert got a job on Park Avenue at a tailor delivering handmade shirts to wealthy clients. During his breaks, he'd head to the library to check out books as well as opera, folk and classical music that he'd play for his siblings at bedtime to lend some peace to the tumult at home.

When he turned eighteen, Robert graduated from high school and worked six months until he saved up a thousand dollars. Then he quit his job, bought a one-way ticket to France on the *Queen Mary* and took off. He lived in Paris for one year and three months, studying the language at two different schools until his money ran out. By the time he got back to New York City in June of 1964, he had one dollar in his pocket, but he was fluent in French. With that dollar, he caught a cab ride home, where he was told there was no more room in the bed for him. From then on, he was on his own.

Robert Witty had found Guinan's by accident. One night in the early seventies, he was driving to a party in the next town of Cold Spring when he missed a turn and ended up down on Garrison's Landing. "I drove down that long and winding road and saw the river, and I was in awe," he tells me. "I knew at that moment I wanted to move here."

I know what you mean, I tell him.

⌒

NOW THAT I'M not at work, Robert and I spend mornings together at the store helping John. Sometimes we'll find reruns of *M*A*S*H* and watch them over and over. Robert has a dog too, Chelsea, a small white spitz mix whose coat is as rough and tangled as his owner's hair. Robert is the calmest person I've ever met. Every morning he gets up and takes a walk in the woods, photographing the snow and ice with an old 35 mm camera. One day we hike down by the river together and take pictures of the ice floes.

"Look at these shadows," he says, pointing to where the sunlight creeps in between two logs. I move around him and shoot a couple of frames. We experiment with the shutter speed and aperture a bit, shooting each other, shooting ourselves and shooting the branches of trees reflected low into the water. Then we sit there on one of the logs watching a long barge push through the ice floes upriver. We are doing nothing. I cannot remember the last time I just did nothing. And yet I feel like I'm in the middle of everything.

When I mention this to Robert he nods. "You've got to live for today because the whole world could end tomorrow." It would sound like a cliché, except that coming from scraggly Robert, it somehow doesn't.

⌒

JIM'S WATCHING THE Westminster dog show in the bar, annoyed because it's preempting *Walker, Texas Ranger* but making the best of it by giving everyone who passes by a running commentary. An Italian greyhound prances across the green and the announcer explains that the dog was raised on goat's milk.

"Holy shit," Jim says, giving a weak cough. "He's proud of himself. Look at that." When the rest of the toy-size group comes on, Jim's got something to say about each one—"Look at his bloomers . . . would you look at that hair, the way it's flowing. I'll tell you, isn't that nice."

June, another friend of the family's, is also helping out now in the afternoons so John can run errands for the store. Her father is the one who drove Colonel Tom home the first night he rowed over as a cadet. She nods her head in agreement at Jim's assessment of the parade of pooches. June has a ponytail perched atop her head, and when she nods, the ponytail bobs up and down to emphasize her point. A Manchester terrier trots by. "Hey, cool," says June, ponytail bobbing. "Cool," echoes Jim.

John gets back and joins us just as a tiny Pomeranian steps up. "Hell, that looks like something your dog, Dolly, coughed up," he says. "It's just a big hairball." I laugh and stroke Dolly, who is lying at my feet.

Suddenly Jim gets up from the bar and heads into the living room, limping severely. John follows him in.

"I want to see your foot," he tells his father.

"It's fine," Jim says.

"Let me see it," John says. Jim reluctantly takes off his shoe and sock. The ankle and leg are bright red and his white sock is covered in brown stains.

"Jesus," John says, clearly unnerved. "When did that start?"

"Yesterday morning," Jim says.

"You should have said something to the doctor," John tells him. "You were there yesterday for that cough."

"They'd have put me in the hospital," Jim tells him matter-of-factly.

John sighs and begins to clean the wound with saline solution, wrapping it with gauze and not speaking. His face is bright red.

Does it sting? I ask Jim.

"No," he says softly and pulls the red-and-green-checked blanket around his shoulders.

"I'm going to call the doctor," John says.

The Pekingese wins.

*LATER THAT NIGHT around 6:30 P.M. my phone rang at home. It
was John. "Hey, kid," he said, his voice tired. "We just got back from the
doctor."*

You were gone all this time? I asked.

"Yeah. It looks like Dad may have another infection in his foot."

Oh no, I said, isn't that what put him in the hospital before?

*"Yeah, that's when he lost his toe," John said. "Anyway, you left your
gloves here this morning."*

I'm really sorry about your dad, I told him.

"It's okay," he replied in a voice that suggested otherwise.

winter storms

Nothing should change anywhere.

Just before Valentine's.

Jane doesn't want to work Friday night because she has a date. I offer to do it instead. I'll be alone because Jim is now checked into a nursing home where the doctors can intravenously pump antibiotics into him and monitor his progress. An operation cleaned out the infection, but it's going to take weeks, maybe months, of healing time before he can come home.

"So you'll be the only one here," John says. He and Margaret are walking me around the store, showing me light switches, the alarm, how to tally the money at the end of the night.

When they're finished with the instructions, John presses something cold into my hand. "Here, put this on your chain."

In my palm lies a key to Guinan's.

I look at Margaret questioningly.

"Yeah, well, if anything's missing, we'll know who to come looking for," she says. But the detective is smiling.

THE ARGUMENT STARTS from nowhere.

It's late on Valentine's Day, around 10:15 P.M., and there are only two guys left. Kathryn's come down for moral support.

"So, I think there should be a movie night here," the clean-cut guy says. I vaguely recognize him; know he used to be a regular here but doesn't drink alcohol anymore. He says it to Kathryn, who leans against the fireplace sipping her Harp and listening patiently. "You do it simultaneously. Show the same movie here and down at the depot theater. And serve wine here before the movie starts. It'll bring in a whole new crowd to Guinan's." He takes a sip from his Lipton Iced Tea, $1.50 a bottle.

I'm counting dollar bills out of the box, adding the night's proceeds, which aren't particularly staggering, when the other guy, the dark-haired one I secretly call the Count, sort of growls from his end of the bar.

"That's the stupidest fucking thing I've ever heard—movie night." He rolls his heavy-lidded eyes and leans deeper into the counter. Tonight he's drinking Budweiser and has been through quite a few. The latest one is curled in his fist like a scepter. After all this time, I still don't know much about the Count but feel like he loves this place with a kind of fierce loyalty that extends to the rest of us, even if he doesn't tell us so. He mostly keeps to himself, sipping his beer, smoking and watching the river— although sometimes, like now, he will unexpectedly pick up on a conversation out of nowhere.

The Count doesn't look at the clean-cut guy when he speaks. Just kind of repeats the last part of the guy's statement with a disdainful tone. "Movie night."

The clean-cut guy blinks. He's wearing a tight gray turtleneck sweater and is boyish in a way that contrasts completely with the Count's brooding good looks. "It's stupid?" the clean-cut guy asks, tightly controlling his voice, which is probably easier since he's drinking iced tea. "Okay, so why is it stupid?"

" 'Cause it just is. It's bullshit," the Count calls over his shoulder, still not turning around. I see Kathryn stifle a smile. "This is a bar," he continues. "Who's going to watch a movie?"

"A lot of people might," the other guy says, undeterred. "A lot of new people in the community. People who don't come in here because they think it's just a place for drunks."

The dark one swigs back his beer and then answers dryly, matching the clean-cut guy's measured tone precisely. "Yeah, well, that's what you do in a bar. You serve the drunks."

"Well, maybe there aren't enough drunks around here anymore," the other man responds pointedly. The line hangs in the air, almost like the fate of Guinan's itself. Kathryn isn't smiling anymore. I lean against the counter and feel the coolness of the red metal cooler pushing through my soft jeans.

"This place should stay just as Jimmy wants it," the Count says. "Nothing should change about it." He looks around. "Where is Jimmy? Is he here? Let's ask him."

Jim's in the hospital, I remind him.

"Yeah," he says. "Well, that sucks." The Count fires up another cigarette, letting his scarf fall across his black fleece vest. He's done conversing. He looks at me instead, energy vibrating off his body. His eyes, though, they are almost sad.

"Nothing should change here," he tells me. "Nothing should change anywhere. Ever." Then he disconnects himself from the bar, picks up the iron fire poker and begins breaking up the embers.

BUT EVERYTHING IS changing. Starting with the fact that Jim is gone. The doctors discover that he's anemic. They find traces of blood where they shouldn't and do a colonoscopy. The test is clear, which is good news. Still, he can't come home.

Meantime, Ed Preusser's grandmother——his last living grandparent——is dying in Florida of bone cancer and heart disease. My friend talks about it at the bar, his voice shaky and eyes red. His grandmother Marie is a strong woman, he says, having buried a husband who died of cancer when he was only fifty. Ed tells me little details, how she was a naval nurse, how her youngest sister was diagnosed with terminal leukemia at age nineteen, how he's the closest to her of all the grandchildren. He spills these memories as if he's scared they might disappear altogether otherwise. "When Grandma goes," he tells me, not even pretending to be a tough guy, "I'm going to be a fucking mess. Too many people I love, just gone."

Then a snowstorm slams the East Coast on President's Day, a gift of two low-pressure centers of warm moist air traveling up from the south that went haywire when they met with the frigid air flowing from a high-pressure system in Canada. The snow experts get out their rulers. By late day, the counts are impressive: Washington, D.C., 15 inches. Philadelphia, 19 inches. New York City, buried and asleep for once under almost 20 inches.

The snow is just beginning to fall in Garrison when John arrives at the store at 5 A.M. as usual. He puts out the newspapers. Wraps the doughnuts. Butters the rolls. Makes the coffee. And then he waits. By 7 A.M., he's had one customer. By 9:45 A.M., there have only been five. His wife calls——"Come home, before you can't." Driving gets treacherous atop the mountain where they live. He promises he'll leave soon, but not just yet——maybe around noon. You never know who might need something. And everything else is closed. His eyelids get heavy. His throat is sore. So he goes outside to shovel and work up a sweat. An hour later, the walk is covered, so he shovels again. A diehard runner comes in for his *USA Today* and *New York Times*. "You're the only thing open around here," he tells John gratefully, his eyelashes frozen, black tights stiff with ice.

After he leaves, John adds another log to the fire and pours himself a cup of coffee.

⌇

THE MEDIA DUBS it the Blizzard of 2003. Winds gust to 55 mph. The storm, 750 miles wide on satellite photos, buries New York City. Cars are visible on the streets only by their antennas. Nearly all flights are canceled out of the three New York area airports. A fifty-five-year-old Bronx man dies of carbon monoxide poisoning while warming up his car because the tailpipe was blocked with snow. In West Virginia, thirteen poultry houses will eventually buckle under the snow's weight, killing some 325,000 chickens and turkeys. A group of gamblers stuck in Atlantic City eventually see the odds turn against them and wire to Western Union to replenish their coffers. The storm is a nonstop news show. Politicians go on TV in ski jackets and promise order in the cold chaos. New York City's mayor, Michael Bloomberg, estimates a cleanup cost of twenty million dollars—or approximately one million per inch of snow. The snowfall is no match for the 26.4 inches in 1947, but still, history book stuff now. People stop complaining. In the city they start cross-country skiing down the avenues. They forget that tomorrow is a workday. Instead, they smile and build snowmen, moving stiffly down sidewalks in thick mummified bundles of fleece. It's a relief, this reprieve from the news of terrorism, our sorry economy and nuclear missile development in North Korea. Everyone's happy to let Mother Nature keep all the gloom at bay, at least for a few more hours.

⌇

IN GARRISON, THE snow falls until decks shudder with the weight and walkways stay stubbornly white no matter how often they are shoveled. By midafternoon, only the bright yellow and orange snowplows dare take to the streets. Their coffee-drugged drivers claim turf, one driveway after another, the inside of their truck cabs strewn with white Styrofoam cups stained with the previous hour's caffeine high.

A few West Point cadets come into Guinan's hoping to catch a train. Each time one arrives, John stays a little longer, just in case another one comes. Finally, at 3:30 P.M., he shovels his dad's walk one last time, empties out the fresh pot of coffee he just made for the last straggler and turns out the lights in the store. He brushes the snow off his truck windshield, turns on the engine and slowly maneuvers up the landing toward home.

OUR PHONE RINGS early the next morning. Or at least it feels early. *My mother,* I think.

"Hey," the voice is low, flat. "It's John."

Hey, I say, trying to focus.

"I just wanted to tell you that I can't get down the mountain. So I won't be there if you show up at the store."

What time is it? I ask, reaching for my watch.

"It's 6:30 A.M."

I think, 6:30 A.M. And the store is closed. *I have a key now.* The thought competes roughly with the warm bed for a moment. Finally I ask him: Do you want me to go open up?

"That's totally up to you."

THE PLOWED STREETS were thin black lines between perfect white fields. A couple of kids zipped down the hills of Nazareth, their tiny sleds whisking through a train of powder until they collapsed, screaming, in heaps at the bottom.

One or two brave commuters were standing outside the door of the store when I pulled up. I hurried to unlock the door with my key, stepping over the piles of newspapers in front.

Inside I turned on the lights, thought for a minute, and then began the workday: coffeepot on, alarm off, butter out to soften. . . .

moving on

I'm tired of seeing it here flat and run-down . . . .

Late February 2003.

The nursing home where Jim is recovering is a nondescript low-rise brick and tan building that sprouts from the earth like a modest mound of dirt. Its individual wings fan around the building, fingers with little satellite dish warts stuck on them. When John and I pull up in his green truck, most of the blinds are closed, a nod to a dreary view that never changes.

We walk down the hallway searching for his father's room. There are old men and women slumped in wheelchairs lining the corridors, their glassy eyes gazing up at us, bits of saliva collecting in the corners of their mouths. One woman clutches a baby doll. Another, her head bent at a 90-degree angle to her body, strokes the matted fur of a stuffed brown teddy bear. About six are packed around the nurses' station. Grandmothers, fathers, businessmen, lovers are now reduced to a mob of white and yellowing flesh waiting for the inevitable. John looks straight ahead.

We find room number 17 with "J. Guinan" tacked to the outside. Jim is looking out the window, or he would be if the shades weren't closed, when we step into the doorway. For a moment, he looks at me confused—I'm out of context here—and then blinks, focusing.

Hey, Jim, I say softly, crossing the room to his bedside.

He brightens. "Well, HELLLOO there, luv." He smiles with recognition. "How are you now?"

"Hi, Dad," John says, coming in behind me. He awkwardly shakes his father's hand. "I brought you some newspapers." He hands him the New York *Daily News,* the *Journal-News* and a boating magazine.

"Margaret already brought me those," Jim says. "She just left. Stayed for a while. Found a spot of blood on my foot. That Margaret, jeez, she'd smell a cat if it'd been gone for three days. Nothing gets past her. She's been here to see me every day like clockwork." He winks at me and chuckles.

John takes back the newspapers and sets them on his lap. "How are you feeling?"

"Fine, fine," Jim insists, making an effort to sit up straighter in the bed. "Feeling much better." He's wearing a green plaid shirt and khaki pants, and his forty-year-old gold golf-club-bag charm swings against his chest. The right leg is bandaged, from a rash, Jim tells us. The left leg is swollen and bruised from the infection and operation. He's got a clear line feeding antibiotics into his arm.

I settle onto the spare bed next to Jim while John begins catching his father up on the news—"Goat Path was closed, traffic was backed up all down 9-D . . . Tip had his operation . . . Ken Anderson's leg is doing better. . . ." All the while he talks, John nervously taps his feet.

There are no pictures on the wall. The paint is mismatched pastels. The RCA TV only gets three channels, Jim reports. No *Walker* at all. There are a couple of greeting cards tacked to the wall above Jim's head, leftovers from a previous patient that no one bothered to remove. Cartoons blare from the room across the way. Out in the hall, a resident yells "TOOT TOOT" and goes scooting by in his wheelchair.

I wonder if we can just grab Jim and run.

⁓

JOHN AND I stay until dinnertime. When a busty nurse comes in to get Jim, the Irishman perks up a bit. "Hello there," he calls flirtatiously. "Did you miss me?" John rolls his eyes, and I laugh.

The nurse smiles. "We're all debating who gets to marry him." She looks at John. "How do you know Jim?"

"He's my father," John says loudly, taking the handles of his father's wheelchair and pushing him into the hallway.

The dining room is filled with wheelchairs circled like wagons around a table. We're the only visitors as far as I can tell, and the rest of the residents look at John and me longingly as we wheel his father over to an empty spot. There's some too-cheery Irish music playing and special diets being served everywhere. John and I linger out of the way by the open door for a few minutes, watching his father across the room.

"Should we go?" John asks.

I nod, feeling like a kid again. This place terrifies me, with its scent of inevitability and demise. John turns and starts down the hallway. Wait, I tell him.

I dart back inside the dining hall and slip over to Jim. We'll see you soon, I tell him, kissing him on the cheek.

⁓

ON THE WAY home we stop to visit Barbara Prescott, who's recuperating in a nearby facility after having a leg amputated because her own infection from diabetes refused to heal. It's a stark reminder of the fate Jim has escaped time and again.

As usual, Barbara is full of fire, sitting up in her wheelchair furiously crocheting what looks to be a blue baby's blanket.

"Who's that for?" John asks, leaning down to give her a hug.

"No one in particular," she says. "But there's always one that will come along."

She's upbeat, likes her new prosthetic, which the doctors are fine-tuning. They tell her she's got three weeks before she comes home.

"And I'm going then," she tells us, tossing the blue yarn out vigorously across her lap, "come hell or high water. Three weeks is it. Even if I have to push myself out of here. I'm getting stronger wheeling this thing around anyway. . . ."

THE NEXT AFTERNOON John puts on music at the bar. It's the first time he's done it since I've been coming here. He turns off the TV—silencing the incessant weather reports, talk shows and *Little House on the Prairie* reruns—grabs a CD of Irish tunes and turns up the volume until the fiddles and guitars sneak through every corner of the store.

Then John goes outside. When he doesn't come back after a while, I walk out and find him furiously pumping air into the tires of the staff car, his breath coming hard and foggy in the cold air.

"I'm tired of seeing it sitting here flat and run-down like this," he says, not looking up.

I GET AN e-mail from our landlords living in Hawaii, telling us that while they appreciate our need to continue assessing the lease, they want to rent the house out for the summer—which around here can generate as much revenue as a year-round contract.

In other words, time's up.

After reading the note several times, I call Kathryn with a sinking feeling in my stomach.

There's a problem, I tell her.

THE NEXT TWENTY-FOUR hours blur together. Late in the afternoon, I take Dolly out for a walk in the fields. It's an over-cast day and snow still blankets the ground. I watch her hop through the snowbanks, weary of being cooped up so long. The waterfall along the creek is frozen solid into long shoots of dangling ice fat as walrus teeth. I lie flat on the ground, letting the cold seep through my jeans, and stare up at the sky.

I tell myself it'll be okay, that eventually we'd be going back anyway. I tell myself that I can still come up for day visits and see Jim and John and Fitz and Ed. That nothing will change, and we can always return here later to look for a weekend home. But even as I assure myself, I know the truth. That once we're gone, once there's no real reason to be here any longer, it'll get harder and harder to come back. As life resumes full-time in the city, it'll happen less and less. Until one day, when I finally do get around to coming back, this whole moment with Guinan's and everyone there may be gone.

My cell phone rings. I'd forgotten it was in my pocket. Fumbling with my gloves, I manage to dig it out of my vest just before it goes to voice mail.

Hello, I say, not recognizing the telephone number.

"It's Melissa," she says, and hearing the real estate agent's voice I sit up. Yesterday, I'd called her in a panic to ask about the immediate short-term rental situation out here, which she's now telling me is pretty bleak. "Look," the agent says, "I'm not coming up with much on that front. But I just got a call. Remember that two-story place I showed you for sale a while back? The one with the stone wall you insisted was from the Revolutionary War?" she adds wryly.

I remembered: missing window, pink walls, crumbling chimney. Despite its flaws, though, there had been something that drew me there. The old wooden gates, it turned out, had been crafted by Mary Ellen's husband, Tony, years ago. It was only a short drive to Guinan's. And there was plenty of room in case, let's say, there was ever a third person and that person required a crib or swing set. The owner had even been negotiable on price, but I'd hesitated—still couldn't make that leap—and someone else had snapped it up.

"Well," Melissa says, "my deal on it just fell apart. You'd have to move fast, but if you're interested, it's available."

I HELD THE telephone to my ear, listening to the real estate agent go on. Dolly poked me with a stick in her mouth, wanting to play. I sat up and hurled it toward the frozen stream with my free hand.

"Live for today," Robert had said, "because everything could end to-morrow." He was right, of course. There were no guarantees about any-thing, even an ordinary Tuesday morning in September. No matter how much I planned and rushed around, the future wasn't totally mine to control. Someday, no matter how much this town loved him, Jim Guinan was going to die. And John and Margaret at any moment could say "Enough," and close the doors to Guinan's forever. But the thing was, they were all still around right now. And so was I.

The gray cloud cover moved briskly across the horizon on its way to somewhere else. I watched it leave and knew that this decision was no longer just about where to live, it was about how to live.

How would I live starting today?

Staring up at the barren slope of the North Redoubt, I found my answer.

the high holy day

But what they don't tell you, is what happens when you live.

Oh, this is not good."

Startled, I jump and drop the armful of dirty curtains I've just pulled off the living room windows of the new house. Walter has snuck up behind me from the basement, where he has been prowling about for the last few hours making a list of things that are broken, worn down, decaying or "just not right." Now he's peering at my walls with the concentration of an annoyed librarian.

What now? I ask wearily, stepping around the pile of smelly material, saturated from years of the prior owner's smoking inside.

"Come here," he commands. "Look at all these nail pops, one, two, three, four . . . my GOD, there're dozens I can see right here." He flaps his arm, pointing wildly up and down the garish pink-hued wall.

What's a nail pop? I say, peering under his arm and wiping a line of sweat off my forehead.

"This," he says, pointing to a little bulge of paint. "The nails holding the Sheetrock to the studs are loosening and starting to pop through." He flicks the bulge with his fingertip, which makes a fleck of paint pop off, revealing a rusty nail head underneath.

Stop, I tell him. You're ruining the wall. I'll just pound them in and paint over everything and it will be fine.

"No no *no no* NO NO NO," he says, voice pitch rising. "They'll all just pop back out and then you will waste all that expensive Pratt & Lambert paint you bought. And why you bought green, I don't know. Linen White, that's what you should be using. It's nice and soothing and premixed. You'd save a lot of money."

So what should I do about the nail pops, Walter? I ask, trying to make him focus.

He takes a deep breath. "Well, first you have to scrape out the Sheetrock and bare each nail head with a five-in-one tool," he says. "Then you tap around the dent with a hammer, nice and easy, to smooth out the edges. Then what I would do is get a screw and use a power screwdriver to drill it in so the screw head overlaps the loose nail and holds it tight. Then you put joint compound on, let it dry and sand it down. Do two more coats, each one going a *lit-tle* wider, sanding down in between. Then vacuum the wall with a wet/dry vacuum to get off any last dust.

"Do that with every nail pop," he continues dramatically, "and you've got hundreds from what I can see. And THEN you are ready to paint."

Walter looks at me. I look back at him, not speaking. We stare at each other like this for a few moments. "Questions?" he demands finally.

I clear my throat. Umh, that sounds like a lot of work, I say.

"It's the RIGHT way to do it," he says. "Haven't I taught you anything? Patience, patience, patience."

Well, I don't have one of those five-in-dime things you mentioned, I tell him.

"It's a five-in-ONE, and you can borrow mine," he tells me back.

And I don't have a drill either, I continue.

Big sigh. "You can borrow that too."

The wet/dry——

"*That* too." He waits, tapping his white-sneakered foot impatiently. "Anything else ?"

Yes, I say, bending back down to pick up the curtains. What's joint compound?

"Oh God," he moans.

⸺⸺⸺

I DON'T SUPPOSE we were the first, or the last, two people to discover midway down the road that our hearts led us to different spots when it came to home. City or country, East Coast versus West Coast—where to settle can be a powerful joiner, or divider, depending on timing.

Even if she didn't plan to bury herself in joint compound, Kathryn did me a far greater favor by accepting that this detour our lives had taken was, for now at least, a permanent path. We were still going back to New York City, but instead of putting everything into one big life in a single place, we would attempt to live much smaller, scaled-down lives in two spots.

While we looked for apartments, the Johnson couple that had loaned us their city place right after September 11 invited us to stay with them a few nights each week. Asking for nothing in return, they cleared out one of their grown son's bedrooms and gave us both a set of keys and we all merged into a makeshift family. It gave Kathryn a break from the commute and let us both reconnect slowly with our old world.

Sometimes it seemed impossible that anything good could have grown out of a day like September 11. But I remembered the lines in Frank the preacher's book. About how after the disaster's pain and suffering spread through the world like ripples on a pond, the healing would travel back toward the pond's center from the shore.

And so with help from so many unforeseen places—the Johnsons, the Guinans, Walter and Jos—we made our way back.

THE BAR IS in full countdown mode to Saint Patrick's Day.

When she was alive, Peg always cooked enough to feed the entire town for free on this day, and the tradition continues. John orders sixty pounds of corned beef and several dozen heads of cabbage. Margaret buys enough spuds and plates to feed three hundred. She also orders a box of Irish soda bread and puts it on the counter to sell, although I see her give away more than she charges for. Robert hangs up the green paper shamrocks and paraphernalia offered by the Budweiser and Guinness dealers. There's a box of pins with little clovers that light up to dole out on the big day. And extra Harp ordered by the case.

With Margaret's help John makes plans to break his father out of the nursing home for the seventeenth. The healing is going slowly and they've just gotten word that Jim may be there for sixteen weeks.

". . . And frankly I wouldn't mind if he does have to stay," John says moodily. It's three days before the High Holy Day, as he calls Saint Patrick's, and he and June with the bobbing ponytail are sitting inside the bar with a couple of other guys. John's had a few Harps and is getting wound up. "It'll keep him from walking around. And if he'd have done what he was *supposed* to do and kept off his feet, he wouldn't be in this position."

I see June flinch out of the corner of my eye. She's very loyal to Jim. "Oh, please," she tells John, tossing her head with the ponytail flipping around on top. "Your father never got up unless it was to go wee in the bathroom. What did you want to do? Bring a bucket to his chair?"

"You didn't see him in the morning, always up and down," John shoots back. He seems mad, but I know this is just his way of coping. Putting blame on his father gives some rhyme and reason to the madness of his illness and makes his dad less of a victim.

"I think John probably knows his dad best," says one of the guys at the bar to June, making the mistake of getting involved.

"I know Mr. Guinan pretty well," June snaps back. The fellow turns red and looks out the window.

John changes the subject and tells June to change the channel. "But not to *Little Shithouse on the Prairie*," he says, taking a final dig at one of June's favorite shows.

She turns it to *M*A*S*H*.

CHRISTINE FLIES UP from Florida to help out for a week. One morning the *New York Times* delivery is short. She writes up the Reserved stack first, putting the remaining copies in the window. They go quickly, and so when a woman, a stranger, comes in late morning looking for the paper, every issue is already gone.

Christine apologizes but says she can't sell from the Reserved pile. The woman stares at her for a moment and then says coolly, "Oh, and I thought this was a *paper* store."

Maybe she's tired, and maybe she's stressed about her father, but the words sting Jim and Peg's youngest child. Because when Fitz comes in not long after, she tells him what happened and starts to cry.

"How could she think Guinan's was just a paper store?" she asks the former federal marshal. "How can she not know what she's standing in?" Fitz calls the woman an idiot and a few other unsavory things and tells Christine not to worry. But the incident stays with her, because the woman has inadvertently voiced what Christine fears is the truth—which is that everything her mom and dad built is slipping away, and one day no one will remember that Guinan's was so much more than a paper store.

MEANTIME, JOHN PUTS the finishing touches on his proposal to the Station Plaza real estate group. It's not exactly business school material, but his ideas and dreams are there. Everyone is feeling optimistic until one morning I come into the store and find John telling someone how he just yelled at one of the board members about the parking situation outside the store.

I know John's under pressure, but I'm annoyed. Your timing stinks, I tell him.

"But they should DO something," he hollers, his face reddening with that Guinan temper. Then he gives me a five-minute discourse on how people abuse the parking outside the store and how he's fed up with it and how the board should fix it.

I listen until he's done. Then I say as bluntly as possible: You've got to hold your temper. And if that board member comes in, you should apologize.

"But they—"

I put a hand on his shoulder and squeeze. Please, I say. For all of us?

"Okay, okay, I know it. Fine. You're right," he says, sighing. He fiddles with the bread knife he's holding, looking down. He opens his mouth, closes it, then plunges ahead anyway.

"It just used to be that whenever there was a problem on the landing, Dad would just call up the General. And he'd come down and there would be tea and they'd work it out and shake hands and that meant something."

He looks at me, eyes reddening. "Now you get the feeling that if you shake hands with someone and turn around, they're just waiting to stab you in the back."

⌒

HE SUBMITS HIS proposal. Soon the board responds: They like his ideas, but could he give them some more firm numbers, such as the cost of the renovations he's proposing and what he expects of them? And a time frame on when a rent increase might be appropriate further down the road if they do make investments in the store?

This is all reasonable, I tell him. Right? You can get this.

"Yeah," he says, sounding unsure. "But I don't want to get my hopes up. I don't want to count on anything."

Because you've been disappointed before about the store? I ask quietly.

He doesn't say anything. He doesn't have to. The answer already fills the room.

THE HIGH HOLY DAY.

Cooking gets started in the morning and so does the drinking. John lumbers around with a Harp in hand directing the placement of Sterno fuel cans and checking in on his friend Stevie, who's directing the food preparation. Folks start drifting in just after noon, piling their plates high with potatoes, corned beef, salad, hot dogs, pasta—more food than they could possibly eat, except that they somehow eat it all and end up going back for seconds. The beer bottles flow out of the red Coca-Cola cooler and John keeps returning to the cellar to haul up more boxes. Even people who don't typically come to Guinan's stop in; a free meal is hard to pass up, after all. And by midafternoon, they're lined up in the store, leaning against coolers, swapping gossip and trying to juggle their plates and bottles without spilling.

Jim gets sprung from the nursing home just in time for dinner with everyone back at the bar. Margaret helps set him up on the couch where he can prop up his foot. He looks tired and gaunt from having spent so many consecutive days indoors. His face is ashen and the blood vessels on his nose are broken and bright red. One by one the party from the next room comes to pay their respects. He nods, welcomes them as if it's any other year. His old friend Tip of the Fearsome Foursome settles down in the green recliner next to the couch and it's clear he's not going anywhere for a while. Eventually I join them on the couch with a plate of corned beef and cabbage.

"Ah, it's wonderful to be home, luv," he says to me.

I nod. And neither of us brings up the reality. Which is that he has to go back in a few hours.

"Hey, Jimmy," a voice calls from the doorway. It's Old Mike, grinning from ear to ear.

"MIKEY," Jim calls, the yellowish curtains ruffling behind him with a sudden burst of breeze through the cracked window.

AT 7:45 P.M., everyone packs into the little bar, still swilling beers and every so often turning an anxious eye up toward the TV, which is flanked by the paper shamrocks Robert hung earlier. Typically, the set would be shut off, replaced instead by Irish tunes or perhaps a live bagpipe or fiddle performance from some of the Irish Night musicians. But not tonight. Tonight, the TV is on waiting for President George W. Bush to address the nation in prime time and tell us that in forty-eight hours, unless Saddam Hussein and his sons leave Iraq, we will be at war.

Kathryn and I are roaming about separately, squeezing between bodies and trying to carve out a space to eat. Alone for a moment, I look around and am struck by the sensation that everywhere I turn, there are faces I know: the Preussers, Mary Ellen, Old Mike, Jane, Dan and of course . . .

"Heh, heh, *heh*." As usual, I hear Fitz before I see him. He's wearing a green baseball cap and is happy and drunk, having been on the wagon since Lent. It's a routine he started back in the early 1980s with a bunch of guys—ex-marines and law enforcement types—to clean up and clear out their systems. Typically they'd all meet at a bar in Manhattan on Mardi Gras, the Tuesday before Ash Wednesday, to tie one on. Next day, the start of Lent, they went cold turkey on the alcohol, breaking their sobriety for only two occasions: one, if a friend visited from more than a hundred miles away (sometimes there were a *lot* of houseguests during this period), and two, for the High Holy Day.

Tonight Fitz is clearly making up for lost time. His eyes are red and voice slightly slurred. But he's in a good mood, putting his arm around his wife—a pretty, regal-looking woman with posture so erect it makes me stop slouching immediately. Fitz's daughters mill around him, occasionally wailing, "DAAD," when he introduces one as "my smartest daughter by far," or "my prettiest daughter, without a doubt."

He edges over to talk. How are you feeling? I ask him, shouting slightly above the din.

"Fine," he says, but doesn't offer much else. When I give him a skeptical look, he pulls me aside in the noisy crowd and leans down close so I can see the prickly points of his mustache. "Hey," he says, "so I may have to sell my house."

What are you talking about? I ask him. I know things are tight because he hasn't been able to work steadily since he got sick. But this is news to me about the house.

"Hey, you know, I've got another daughter coming up to college, and you can't treat one differently than you did the others. And the college tuition, it's just out of control. My son's tuition the last year at Fordham was $17,500. It took $100,000 to get my daughter through college at the University of Rhode Island. Even one of the local colleges is $16,000 a year now."

The president's speech has started. We both look up for a moment and listen to Bush lay out his case for war: ". . . *Intelligence gathered by this and other governments leaves no doubt that the Iraq regime continues to possess and conceal some of the most lethal weapons ever devised. . . .*"

Fitz sips his beer and then gives me the old elbow. I turn my attention back to him.

"Hey, listen to me. My wife and I," he says, "we never fought. Not about money. Never. Now that's all we ever fight about."

What about a home equity loan? I ask, trying to think of something.

He smiles. "Already done that."

" . . . *Saddam Hussein and his sons must leave Iraq within forty-eight hours. Their refusal to do so will result in military conflict, commenced at a time of our choosing. . . .*"

I lean my head down, upset at this news about Fitz. He takes a finger and lifts my chin back up so my eyes will meet his. The gesture is not cruel, but it is firm. "The thing is, they thought I was dead a few years back from all this stuff I've got. They called in my wife at the hospital and told her this. And suddenly you start thinking to yourself, 'Okay, I've got two Christmases left, three birthdays with my children, this many sunsets. I bought a ton of life insurance. And everything, everything they tell you about appreciating what's around you is true. You savor a walk outside or a glass of milk. You really do."

He elbows me again.

"But what they don't tell you is what happens when you live. What happens when you don't die and suddenly you have to go back to the land of the living and pick up where you left off? What then?"

" . . . *The tyrant will soon be gone. The day of your liberation is near. . . .*"

I'm quiet for a moment, then put a hand on his arm, not sure what to say.

He shakes his head. "I don't know why I'm telling you all this. I've been drinking. Maybe it's because we're about to go to war, and I'm thinking about all those boys."

We both glance back up at the TV. A few other people are also still straining to listen above the din, though most have given up and gone back to eating and drinking.

Fitz takes the last swallow of his beer and looks over at his daughters chatting happily by the fireplace. "You can't treat any of your children differently," he says, more to himself than to anyone. Then he points toward his daughters, the finger stiff and curved.

"Hey," he calls to them, grinning. "Now which one of you is the smartest?"

"Americans understand the costs of conflict because we have paid them in the past. War has no certainty, except the certainty of sacrifice. . . ."

ALMOST NO ONE *saw Jim leave to go back to the nursing home. Around 8:30 P.M., Margaret helped him into the wheelchair propped to the side in his living room. Rather than push him through the crowds, she opened up the "quality door" and they left quietly together without any fanfare. On the way out, her father rolled past his set of dusty golf clubs, which leaned against the staircase waiting, like all of us, for him to come home again.*

22

settling down

I believe that's how we choose our homes.

The sign on the outside door of the castle reads:

BEWARE.
WE HAVE LIVE RATTLESNAKES IN THE HOUSE.
THEY ARE OUR PETS.
HOWEVER, IF STARTLED OR PROVOKED,
THEY WILL BITE.

I peer into the foyer uncertainly.

"That sign's just there to deter unwelcome visitors," the man accompanying me says cheerfully. I smile at him, relieved. "Although I've definitely had more than one encounter with a real rattlesnake here," he adds. My smile fades.

It turns out the crazy caretaker at Osborn Castle picks up his weekly *Barron's* magazine at Guinan's and has been a friend of the family for years. After all the big talk around town, it only took a phone call for him to agree to show me his prize. The two dogs I fled from in the fog are now lying quietly at their master's feet, one nibbling ticks off the other.

As sole caretaker, protectorate and occasional resident of Osborn's old estate, Chip Marks doesn't seem so much crazy as just in love. He tells me he's been tending to the castle's needs since 1977 in one form or another. After it passed into the hands of its current owner, an investor who rarely sets foot on the property, Marks stubbornly lived inside year-round to protect his charge. In the winter, he moved his sleeping bag from room to room as the temperatures dropped. With snow blanketing the region, 45 degrees Fahrenheit was often as warm as it got inside, but Marks stayed on, meditating through the cold. Only when the plumbing was ultimately shut down did he finally move into adjacent quarters to sleep, although he spends nearly all his time performing triage on whatever part of the castle is crumbling most severely. In his free time, what little of it there is, he huddles over a desk in the old dining room penning his stories and a collection of philosophies he's named "A Modest Manual for Living on Earth."

Marks is fifty-one years old, yet sinewy and strong as if he were twenty years younger, a virtue he attributes to diet, herbal supplements, the mountain spring water he collects by the gallon in old apple juice containers and decades of manual labor caring for this building and her land.

"I feel very strongly about the castle," Marks says, leading me through the front door and across a Minton tile floor that retains its original majesty despite layers of dust and peeling paint around it. "As a piece of architecture, it grabs the imagination." There are about thirty rooms inside, he tells me, at least nine fireplaces and an equal number of bathrooms. In the old dining room, he points to a hole in the floor. "There used to be a buzzer there, which the lady or man of the house pushed when they wanted the next course served."

We venture out a wooden door onto a wide slate and stone veranda. "But as strongly as I feel about the castle, more important than the building is this actual spot," Marks tells me.

Before us, the Hudson River sprawls out like a thick snake, her treacherous S-bend at World's End more clearly visible than I've ever seen. Spring has taken the Highlands in a paint-by-numbers fashion. First came the bright yellow patches of forsythia, its blossoms appearing almost overnight as splotches of color on the gray landscape. Their arrival sounded the starting gun for the rest of nature, with thousands of daffodils soon poking through the earth and the trees glowing lemon-lime.

The view here at the castle extends for miles, almost maplike, with different bridges, property lines and roads sketched out. Boats inch along the river like little white dots, the hum of their motors climbing above the fields and farmhouses up to where we stand, higher than the tree line, with a view of the entire world, it seems.

I take a deep breath. Is it true, I ask, what they say about Osborn and the magnetic pull of the earth leading him up here?

Marks chuckles. "As far as I know he had dysentery from his years working in the Philippines. One day he drank water from the mountain spring here while on a picnic with his wife-to-be, Virginia. He found the water stilled his system and then purchased the entire mountain so he could own this spring."

I must look a little disappointed because Marks motions me over to the edge of the castle's wall.

"Feel this, though. The energy of this spot. It doesn't matter what was built up here," he tells me, waving an arm across the horizon. "It doesn't matter if it was a castle or shack—it would be a monument. Some spots, like this one, will feel sacred and soothing to anyone because of their tremendous power. Other spots feel comfortable only to individual people. I believe that's how we choose our homes."

A WEEK OR so later I crouch atop the roof of my own little castle, such as it is, struggling to keep vertigo at bay. With me is the just-hired mason, Roger, who describes what he believes is the real problem with the leaking chimney.

"It's going to collapse," he says.

Waving a beefy arm toward the chimney's north face, Roger Chirico points at what from ground level seemed like just a few loose bricks clamoring for a quick coat of stucco. "Here, pull on this," he urges.

Feeling vaguely ill, I give a weak tug to the face of bricks, which immediately detaches from the chimney with a loud crunch. While Roger muses on how the guy who built this place clearly didn't have a clue, I stand helplessly clutching a critical piece of my fireplace, calculating the thousands of dollars in repairs and wondering how I ever expected to survive out here without a well-staffed building maintenance crew.

"Hey," Roger says kindly, bringing me back to my new reality, "don't worry. We'll work something out, and I'll make it better than new."

I look at the deep green valley around me and think about Marks running around with his ax and dogs performing emergency surgery where his beloved castle is hemorrhaging most. My perch is nothing by Osborn's standards, but I can see how the loose stone wall laps around the valley, breaking only for Tony Yannitelli's wooden gates. There's the fenced-in yard, where Dolly lies chewing on a stick. And a sprawl of rugged evergreens lining the dirt driveway—one no longer available in some real estate agent's window. Seeing this spot in the world laid out so plainly, something shifts inside me, and I know I'll do whatever possible to protect it.

WALTER GIVES ME a new book called *Dare to Repair*. "There are some good chapters on electrical wiring," he says, grinning.

To call his bluff, I turn to a section called "Understanding Switches and Receptacles." I read a little bit—it's actually sort of interesting—so later that afternoon I read a little more. And then, just for fun, I ask the Pidala electricians working on the house to show me how to rewire an outlet. The next afternoon I'm in Home Depot and find myself lingering over voltage indicators and wire strippers. Soon I'm back with a bag full of tools and studying installation instructions for a single-pole light switch. Out of curiosity, I attempt to replace the cracked yellow hallway switch with a new white one, not expecting it to actually work. I clip the old wires, reattach them to the new switch, wrap electrical tape around the whole mess and then go downstairs to turn on the power and test my handiwork.

I flip the switch, and . . . the light works. I'm so surprised that I stand there turning the light on and off and on and off until I'm certain it's not a fluke. At which point I decide, well, why not keep going, and soon I'm halfway through replacing all the old switches and outlets—half of which don't even work—across the living room.

I also fix every single nail pop by hand. It consumes weeks, but by the time I'm done, every wall is smooth as silk. After Walter signs off on my progress, he takes me shopping where he picks out the proper Purdy paintbrushes and cleaning tools. Each night after working, I wash and spin-dry the brushes, scrubbing carefully between each bristle while I hear Walter's voice echoing in my head: "If you take care of it, it will take care of you."

If the house is a disaster inside, the outside is a wasteland. The last owner's dogs have killed almost all the grass in the front yard. The bushes have crept across the doorstep, and what looks to have been a rose garden is so overgrown you can't get near it.

One day while I'm inside painting, Walter and Jos drive over and present me with a rake and a shovel that have a red ribbon tied around them. "It's Craftsman," notes Walter. "Guaranteed for life." I nod, and thank them, wondering when I'll ever have time to even think about the yard. But then Jos returns to their car and pulls out her own set of tools and heads to the front yard, where she straps on some knee pads and kneels down in the weeds.

What are you doing? I ask her.

"Well," says my old neighbor, yanking out a handful of wild onions, "we've got to start somewhere."

THE DOCTORS SAY Jim is healing nicely and can come home in a few weeks. Before his father returns, John halfheartedly goes about fulfilling the tenants' association's requests. He gets a couple of estimates for redoing the floors and shoring up the riverbank.

With the weather getting better, however, there are more pressing things to do, he feels, than draft a business plan. Like cleaning out the store shelves, organizing the Archives and trimming branches by the river so his father will have a better view when he returns. He stores more inventory in the cellar to clear the aisles and talks about painting to spruce things up.

I watch him do all of this and am frustrated that he just won't give the board what they want. His sisters and half the morning commuters could help him write a plan. But John's pride is still hurt, not just from what happened with Christine and his father so long ago but also from the old rumors about some folks wanting the place to close. So for now, Jim Guinan's eldest son practices what he knows, or what his father taught him. He gets up at 4 A.M., makes the tired commuters smile and remembers to give Lou-Lou her morning roll.

"If they don't want to put my name on the lease, then fine," John says. "I'm just going to keep showing up until somebody tells me not to."

WHEN JIM ARRIVED back in Garrison, he surveyed all John's little changes—the clean shelves, the migration of the buttered rolls to a new location. For a few days, he and his son navigated around each other, not saying much. Occasionally Jim grumbled under his breath about this or that. And John would roll his eyes or tell his father that this is the way things were being done now and for Christ's sake, would he sit down before he ended up back in the hospital? Then eventually Jim took his place on the pink couch, and John resumed his post behind the counter in the mornings.

It wasn't exactly a ceremonious changing of the guard. But then, ceremony was never really the point of Guinan's anyway.

23

blackout

Dad wouldn't close down.

August 14, 2003.

A freak reverse power surge blacks out much of the Northeast and Midwest, sending eight states into a darkened mess where much modern-day technology is useless—air-conditioning, ATMs, cordless telephones, elevators, trains, computers and traffic lights. Cell phone networks freeze from demand. Stifling hotel rooms across New York City empty, the guests paying $350 a night to camp on the streets using a water bottle or jacket as pillows. With most of mass transit shut down, thousands of subway passengers are trapped in the hot tunnels and forced to evacuate through trapdoors to daylight.

The blackout hits shortly after 4 P.M., in what one utility official calls "a blink-of-the-eye second." I'm in the car driving home from North Carolina, still about two hours away from Garrison. My phone, which has been in a dead zone, starts beeping frantically with messages from my mother, who initially, like much of the country, presumes another terrorist attack. But instead the radio announcers are talking about a lightning strike in Canada and an antiquated power grid as the culprits. By the time I reach home, the highway exits are clogged with commuters outside gas stations. New Jersey's governor suspends all tolls. Offices close and workers head onto the streets to sweat it out with cases of beer. Commuters rent bikes and scooters or start walking and hitchhiking home as far as Connecticut.

It's nearly dark when I pull into Garrison. There's a waning full moon, which reminds me that it's Thursday, Irish Night—or would be if this power mess hadn't happened. I wonder what time Guinan's had to close with no electricity. At home, having failed to hook up a gas tank yet to my generator, I scramble to find candle stubs and my lone flashlight, and quickly empty out the refrigerator, stacking milk and eggs into a cooler with ice. Kathryn is traveling, so after I finish securing the perishables, I find myself standing alone in the living room, staring at the candles and thinking, Now what?

I call Walter and Jos, who tell me to come over. "We're cooking beans and hot dogs on the grill," Jos says. "Wait, hold on, Wally's saying something. What, Walter?" She breaks off, and I can hear Walter in the background giving her instructions. "No, Walter," Jos says, "she does not have to bring her own beans. . . ."

On my way to their house, I pull down to the landing, figuring I'll just check and make sure Jim is okay. The Yannitellis' house is strangely dark, as is David's bookshop and all the other little row houses. I keep an eye out for Lou-Lou and slow to a stop outside Guinan's. It's dark too, and I can't see any movement. Jim's probably asleep, I think.

I'm about to back up and leave when I hear the faint sound of music through my open window. And then singing drifts through the darkness into the car.

I park and walk toward the house. There sit a handful of the musicians, playing in the dark on Jim's front porch as if nothing out of the ordinary had happened. Jack McAndrew, the mustachioed session leader, is strumming away on his guitar resolutely even though there's no way he can see the strings. A couple of guys are listening on one of the old benches from the *Hello, Dolly!* set, their bodies silhouetted against the moonlight, beer bottles lined up beside them. "Hey, you," one of them calls. "Where've you been?"

It's Fitz; he's holding a beer and smiling. Just then Jane emerges from the store, heaves ice out of the chest and onto her shoulders. I follow her inside, where I find John sweating and scattering ice cubes over the cold cuts and ice cream while swigging from a Harp. A single flashlight illuminates the inside of the store with a kind of eerie glow. There's not a big crowd, just a few people coming in and out.

I can't believe you're open, I tell John.

"Dad wouldn't close down," he says, shrugging. "Said everyone was going to show up anyway, and besides, we don't need power to run this place." I glance at the silent old wooden cash register and could swear it looks a little smug.

Jim is outside, in his favorite golf shirt, sipping a Haake-Beck and listening along with everyone else. West Point is actually lit up, with its generators pumping overtime. I slip back into the bar to get a quick beer before dinner, forgetting the flashlight. A customer walks in and sees me hovering in the darkness.

"Can you get me a Heineken?" he asks.

Sure, that's $3.25, I say, and without thinking reach into the left cooler, straight to the back, and in the dark pull out a bottle from the piles of Harp, Guinness, Ballantine and Rolling Rock.

It's a Heineken.

JIM SOLD THE staff car. It left Guinan's in the afternoon by tow after a morning lounging by the river, its square nose pointed toward the Adirondacks. Back in the bar, a few folks toasted the chariot that had ferried Jim and Cliff and all the boys for so long. It's the end of an era, they said.

The "Guinan 1" license plate was reattached to another car—a black 1986 BMW that young Jimmy bought on his last trip home for his dad. Jim liked his new set of wheels, even though he couldn't drive it himself yet because of the clutch and his bad foot. But he was careful to remind everyone that there was nothing wrong with the old staff car.

"She had a lot of life left in her, she did," he insisted.

the reinforcements

My God, look at all the COUPONS.

On October 2003, Jane tells John that she can't help at the store anymore. Her departure isn't a complete shock—she's been hinting that she needs to spend more time with her kids—but when it finally happens, the timing is terrible. John and his wife are scheduled to take their first vacation since the whole saga with his father's foot began. On her last day, the Friday-night crowd drifts in one by one. Ed, Clemson, the Count and then Mary Ellen, who bursts through with a cake and a card that reads: "I believe in a bright and shining tomorrow ahead of you." She does a little dance around the bar and sings the exact verse she's had inscribed on the cake: "Jane, don't go, we all love you so."

John tells his landscaping boss, Lew, that he's got some things to take care of. For a few days he stews about the store until Margaret takes him aside. She tells him to pull himself together—says that he will go on his vacation and that they will find a way to make things work. "We always do," she reminds him.

THEY FIND REINFORCEMENTS none too soon. The first is a cherubic motorcycle mechanic named Andy who used to live just across the tracks with his mom and has recently begun drinking at Guinan's. Andy is a good-natured intellect who is a tenor and sings opera in his spare time. We hear he's got real talent, although the only time most of us have heard him was when he belted out a boozy aria in the bar after someone gave him a dollar. One Friday night, Andy brings in a plate of cucumber finger sandwiches and plays jazz CDs. The parishioners are so happy he's here that no one seems to mind. And well, actually, the sandwiches are pretty good.

The other is a petite artist with a lonesome countenance who works at the local arts center. She has big glasses and a sweet laugh and piles the sandwich meat on thick.

Both she and Andy fit in perfectly.

Meantime, John's own children offer to help, adding a third generation of Guinans to the ranks. His sixteen-year-old son, Casey, has inherited the arithmetic gene and can tally up the orders in his head as fast as his father. After school, he boards the train and rides down to Guinan's, where he hangs out until Andy arrives. His sister, Kelly, a twenty-five-year-old graphic design student, comes in on weekends and does homework while she tends bar.

Everything seems to be set for John to go on vacation. But the Monday before he leaves, I walk in to find him in a horrible mood, snapping at his father and banging the mop around too hard as he scrubs the floor. He has forgotten to take his blood pressure medication that morning and his face is bright red. Tip Dain of the Fearsome Foursome is there watching him and looks worried.

What's wrong? I ask John. When he doesn't answer, I turn to Tip.

"He doesn't have anyone to open the store while he's away," Tip says. "New folks can't do it, and he thinks he may have to cancel the trip."

I think for a moment, and then grab a piece of paper and sketch out a calendar. Okay, I tell John, I'm traveling at the beginning and end of the week, but I can take Tuesday and Wednesday.

He brightens a little. "Well, Murray up the road here can probably do Thursday and Friday," he says. Then he sighs. "But there's still Monday. And I don't know who else I can trust with all the money."

I have an idea, I tell him.

⌒

"FREE," WALTER CROWS, shaking a package of Bic razors. He's just returned from a two-hour trip to the supermarket and is brandishing his receipt high in the air as he dumps six more bags of disposables onto the counter. "And the ice cream was a deal too," he continues, plunking down four pints of Häagen-Dazs. "I doubled up my coupons and the Häagen-Dazs was already on sale, so I made money. God, I'm good."

Jos and I stare at him; I've already filled her in on my plan. "What?" Walter says, looking back and forth between us. "Doesn't anybody care? This is important stuff."

Walter, I say, I need a favor.

⌒

IT TAKES SOME convincing—"Just imagine, you'll get to handle all that money, Walter"—but my former neighbor finally agrees to open up on Monday with the condition that Jos comes too. "In case I get nervous and can't think," he says. And with the last piece of duct tape in place, John goes on his vacation.

Monday morning, Walter and Jos show up at 4:30 A.M. They make the coffee, butter the rolls, wrap the doughnuts and introduce themselves to the tide of commuters. Jos, aside from being a horticulturist, also turns out to be a math whiz, and so she tallies up the orders throughout the morning rush. Walter, meantime, busies himself at the drink cooler, pulling out bottles, inspecting their expiration date and putting each one back so its label faces front and center.

But he really gets into the spirit of things when he steps outside for a smoke and notices the stacks of unsold Sunday newspapers waiting for recycling. He prowls through the piles and gives an excited shout back to Jos: "My God, look at all the COUPONS." By the day's end Walter has gone through every *Daily News* and *New York Times* to amass for himself an arsenal of discounts, including more than a dozen 20-percent-off cards from Linens 'n Things.

I worry he may never leave.

At 11 A.M., the lady with the big glasses and sweet laugh shows up. John's son, Casey, comes down at three and Andy at six. Margaret arrives earlier and stays longer than usual, keeping an eye on us all until she goes off to work. Even though she's "officially" retired from the store, it's hard to tell because she's still here every day cooking for her dad, taking him to doctors' appointments, stocking shelves and making sandwiches.

Since none of us are pros at this yet, on Irish Night it takes a handful of people to hold down the fort in Jane's absence. Clemson the Latin sportscaster comes down to help, as does Mary Ellen. Together they load beer and soda into the coolers, hauling heavy boxes up from the cellar while Andy and I tend bar. Occasionally Jim peeks out and scowls when he sees so many of us scurrying around behind the counters. But eventually he mellows, has a Haake-Beck and even sings "Danny Boy."

We're a little rough around the edges. Still, between Margaret and all of us, the true believers, the chapel survives the week. It isn't exactly how mom and pop did things. But this is the kids' show now. And so far their patchwork is holding.

＊

WHEN HE RETURNED *from vacation, John was wearing a new shirt. It said: "God put me on earth to accomplish a certain number of things. Right now I am so far behind I will never die."*

His eyes were still tired, but there was a look of resolve about him. When the rumors surfaced now about the store closing, John would shake his head. "People are waiting for me to throw up my hands in frustration," he'd say. "But it's not going to happen."

I still didn't know whether he was doing this for himself or from duty to his dad. But then I wasn't so sure the two reasons were all that different anymore.

25

thanksgiving

. . . thank you for helping hold everything together. . . .

November 27, 2003.

It's crisp and cool. The clouds hang low in the sky, wrapping around the Highlands mountain ridges. My parents have driven up from the South, and around 11 A.M., my mom and I head down to Guinan's. Jim is up and about, wearing two regular shoes. His left foot has nearly healed. An unusually cheery Margaret is sipping red wine in the kitchen with John's wife, Mary Jane. As her mom used to, the detective has whipped up a Thanksgiving feast—peas, carrots, turkey, turnips, celery stuffed with cream cheese and chocolate cheesecake.

John stays behind the bar until noon, when Guinan's officially closes for the holiday. After that, all the "elite," as his father dubs us, get to stay on. Once everyone else has cleared out, Jim pulls a twenty from his pocket and buys the bar a round. Old Mike is here. And Fitz and another guy named Sam who belongs to the boat club. Ed Preusser wanders in. The Macy's parade hums from the TV, but you can't look at the screen too long without getting a headache because the picture's been fuzzy ever since a lightning storm hit a few weeks back.

John is in good form, drinking Harp, telling stories. Tonight he will carve the turkey—and is insistent on regaling everyone with his technique in full detail. He's well into the speech when I look up at the doorway and see Governor Pataki standing there in a baseball cap. He sizes up the room and quickly rattles off a round of personal greetings: "Sam, I haven't been getting much advice from you lately, what's wrong?" "Ed, how are your parents?" I've only met Pataki twice, but he remembers my name and then shakes my mother's hand before sitting down next to Jim.

"I haven't been kidnapped in a long time, Jim," the governor says affably, stretching out his long legs. "When's it gonna happen again?"

Jim laughs, but my mom—who's the only newcomer here and always a little jittery above the Mason-Dixon line—looks concerned at this bit of information and elbows me none too subtly, making me bump into Ed on the adjoining stool.

"When I was assemblyman," the governor says, smiling, "I used to come down here to Guinan's all the time. Back then there wasn't this huge security detail hanging around. And one year there was this big parade down in Peekskill and a concert with a great Irish band called the Wolfe Tones."

Jim is nodding and clearly likes this story. "That's right. That's right," he says.

The governor continues. "So I had a bunch of things going on and only wanted to go to the concert for a little while. But Jim here, he gets so excited and says, 'No, no, we're getting a bus and we'll bring you back here to Guinan's after it's over. You have to come back on the bus with us, George,' he tells me."

"Only way to go," Jim chimes in.

"And I said, 'Okay, but, Jim, you gotta promise that as soon as this thing is over, we're coming straight back because I've got a lot of things to do.' And he promises," the governor says, laughing and looking over at Jim.

Jim chuckles. "That I did."

I steal a quick look at my mom, who seems to be relaxing. And the governor continues.

"So the Wolfe Tones were great. But when the concert is over, Jim and Mike over there, they both usher me on the bus just a little too quickly. And as we start to pull out of the parking lot, Jim suddenly goes 'Guess what? We're heading to Brodie's Pub.' And I said, 'Jim, we're supposed to go back to Guinan's.' But he'll have none of it."

Jim shakes his head and pats his knees with delight, his gold golf-club-bag charm swinging from the chain around his neck.

The governor puts a hand on Jim's back: "Well, I guess I didn't protest too hard. So off we go to Brodie's. We walk inside, and now I may be the local politician, but everyone there knows Jim. And lo and behold, the Wolfe Tones show up at the bar and we stay with them drinking and singing until four in the morning, when we finally get back on the bus to come home.

"It's snowing by now," Pataki continues. "Big wet flakes, and the ground is covered, and it's totally dark back here on the landing. Mike offers me a ride up to my house from Guinan's, but I'm not thinking clearly and so I insist on walking up the hill."

Old Mike chuckles and nods affirmatively over his Schaefer at the bar's end.

"Unfortunately," Pataki says, "I am wearing dress shoes at the time, so I keep sliding back down the driveway into the snow, and so by the time I finally hit the front door, I'm soaked."

Jim can't contain himself anymore. He puts a hand on the governor's knee and takes over the story. "His wife, Libby, she calls me the next day and says, 'Well, what did you do to my husband, Jim? He came walking through the door in the middle of the night singing at the top of his lungs, looking like a drowned rat, and told me he'd never had a better time in his whole life.' "

"It's true," the governor says. "I don't remember what I was supposed to do the next morning, but it did not get done well. As I've always said, Jim, anytime you want to trade jobs, I'm ready. I'd have a lot more fun doing this."

Everyone laughs. Then I see John standing alone behind the bar, his turkey speech forgotten amid his father's escapade. The governor must notice too, because after a moment he gets up to leave and walks behind the bar. "John," he says, loudly enough for everyone to hear, including Jim, "thank you for helping hold everything together here. I know it's hard work, and you and your sister are doing a great service to the community."

It's just two sentences. But right now it seems more like a kind of unofficial passing of the torch, at least to anyone paying attention. John smiles. "Thank you, Governor," he says, and looks over at his father for a moment. Then he nods at him inclusively. "We're doing what we can."

"Well, Garrison would not be Garrison without Guinan's," the governor says, turning to leave. "Happy Thanksgiving, everyone."

AFTER PATAKI LEAVES, Fitz decides he likes my mom and her southern accent. Offers to buy her a beer. "Hey, a family that drinks together, stays together," he crows. Margaret brings out hors d'oeuvres for everyone. John resumes his turkey-carving explanation, while Jim sips his Haake-Beck and sits in the midst of us all, his congregation.

Everything has changed, I think looking around, and nothing has changed.

The phone rings, and a chorus of voices sing out predictably: "I'M NOT HERE." Margaret picks it up in the kitchen and then comes out to stand at the door of the bar, hands on hips, in front of Fitz. He's dressed up for dinner, wearing a suit jacket with his federal marshal pin.

"That was your wife," Margaret says. "She said to tell you that your son will be in on the two o'clock train, and then get your ass home."

He laughs. He*eh, heh.* "Now wait, wait. Tell me. Was that what my wife said—'Get your ass home,' or was that the DT talking?"

The detective sidles next to him. "She said, 'Tell him to pick up his son, then kick him out.' I added in the 'get your ass home,' no extra charge."

"He*hee, heeeh.*" He wraps his arm around Margaret's waist, settles back on the stool behind the bar and plunks down his empty Beck's.

"Well, in that case, I'll have one more."

DOLLY RACED AHEAD of me up the wooded path, winding her way along the trail to the top of the North Redoubt where Washington's troops once stood. It was quiet and cold, the landmarks of these last two years hidden in the surrounding landscape. To the left was Osborn's castle; to the right our old house; straight ahead down somewhere in the trees, Guinan's.

The bonus puppy was still young, with legs that quivered at each new smell. I leaned against a rock as she tumbled around until finally coming to sit quietly at my side, her body warm, chest expanding and contracting against my legs.

Together we rested and listened to the wind rustle around the mountain—nothing more. But for that moment at least, I heard it clearly and thought I understood what John had really meant when he said, "Pace yourself, kid. It's a ride, not a race."

home, again

Lou-Lou's at the door.

Christmas Day.

Ken Anderson ambled into the store, prompt as always.

"Ken, how the hell are ya?" John called out.

"My wife, Helen, made you some cookies," Ken replied, plopping the tray down in front of John and reaching for his newspaper.

John picked up the batch, touched by the gesture, and shook his head. "Wow, Ken. Thank you so much. Seriously, that's the nicest gift I've gotten."

Ken tucked his newspaper under one arm and gave a knowing smile. "Don't bullshit me," he said, and then headed out to his little red car, its engine puffing in the cold. "Merry Christmas, everyone," he called.

". . . AND THE storm will extend further and further north overnight," the weatherman droned.

Dan the lawyer stared up at the bar's TV and then noted matter-of-factly to Fitz, who was standing beside him: "He shouldn't say 'extend further and further.'"

Fitz looked at him for a moment and then took the bait. "What, because he shouldn't say 'extend'?"

The lawyer glanced at him smugly. "No, he should say 'farther.'" A couple of folks chuckled.

"Would you just watch the damn weather, Dan," Fitz retorted, getting up to pace as the battle heated up. "You are so anal. I swear, when you go up in your airplane with your list of things to check, you'd better write down 'check your fly' to see if it's unzipped because. . . ."

KATHRYN FOUND US a small, sweet apartment downtown. Like the house, it too needed help, but having trained under Walter's strict standards, we were able to do a lot of the work ourselves. We didn't even need the superintendent—well, not often, anyway.

Across the street was a dilapidated old pub and a few blocks over was the river, which somehow made everything seem not quite so far from our life up north. Because standing on one of the city piers and looking right on a clear summer night, you could see how the Hudson snaked out of sight up toward the Hudson Highlands and on to the Adirondacks. And late at night in the country, when all was still and dark, the sound of the train rumbling by on her shores was a gentle reminder of the bright urban life flourishing down the track. During the week, we took turns commuting between the two places, and while it maybe wasn't the simplest living scenario with all the suitcases and late-night trains, we did our best to make it work.

Like the river that flows two ways, the trick was finding room for both currents.

I ALSO WENT back to work at the *Journal* but didn't stay in the same job. Instead I started writing a column about small businesses, which in a way was kind of like writing about the Guinan's of the world every week. There were no fancy parties or big meetings anymore. My Jil Sander suit sat in a plastic dry-cleaning bag in the back of the closet, and most days I wrote from home, clad in blue jeans, with the bonus puppy at my feet. It was a far cry from the stuff of my paperback novels, but when Ed Preusser's grandmother finally died, I was around to be at the funeral. I was also there to hear Walter's latest coupon coup or Jim's infallible weather predictions. And if John needed extra hands behind the counter or bar, mine were sometimes available. The big game went on without me, and maybe one day, I figured, I'd fight to get back in. But for now the view from the wooden bench outside Guinan's was fine.

The day I changed jobs was Irish Night, and that seemed to be a pretty good omen. The Highland Harper stopped me in Jim's kitchen to hand over a copy of her new CD—*Flights of Fancy*. I thanked her and inquired about Zip the rat.

"He's slowing down a little," she told me solemnly.

I told her I was sorry to hear that.

"But I've still got Ralph," she said, brightening. "He's a one-pound rodent who chases my seventy-five-pound husky. Ralph thinks he's going to medal as an Olympic jumper too. You should see him hopping around the bedroom. . . ."

Later we all packed into the bar, listening to the harpist and other musicians strum their instruments while the trains thundered by outside. Andy the opera singer and John's daughter, Kelly, worked the bar. Mary Ellen danced by herself in the doorway. And Jim got up to sing "Danny Boy." As we listened solemnly, his hymn floated from the bar to the river, from the river to the sea, and then, I imagined, back to the shores of Ireland.

MARCH 17, 2004

The High Holy Day came. Back in Garrison people spilled out of the store, devouring corned beef and cabbage. Jane returned for the evening to help out. Fitz and his wife were leaving as we arrived. Their house was about to go on the market, but I knew that even after they moved, the ex–federal marshal would always find his way back here on this day.

Tonight, though, he was heading out early, having been at Guinan's since 1 P.M. after breaking his own rule and falling off the wagon the night before. "It was the moon," he swore to me, putting his arm around his wife. "I swear it was the moon."

At some point amid the fray, John walked into the kitchen and looked at his father. They'd been bickering about this and that during the day—how many people were coming, when to start the corned beef—and John figured, what the hell? He was tired of always nagging, tired of fighting. So he and his father poured a shot of whiskey together. And for once John wasn't waiting to hear certain words from his father before acting. "I love you, Dad," he said before they toasted.

Maybe that drink wasn't the best thing they could do together. But maybe it wasn't the worst either.

SPRING ARRIVED NONE too soon, plumping the emaciated branches and coaxing out life everywhere in the country—including the weeds again. There were slick flat ones—"plantains," Walter proclaimed when I called him, panicked—that carpeted the yard by the thousands. And dandelions that seemed innocuous enough until suddenly that's all I could see in one section of the yard. One weekend, Kathryn caught me darting about with a lighter setting fire to their seedy, white fluffy heads, a sign that Walter's indoctrination was nearly complete.

"Are you doing what I think you're doing?" she asked, laughing.
I hate 'em, I mumbled to her. Hate 'em.

AND LIKE THE seasons, other things kept changing around us:
A fancy new market opened at the top of the hill serving lattes
and sandwiches with goat cheese. An investor turned Dick's
Alhambra-like castle into million-dollar condominiums. Pabst
Brewing Company stopped making Schaefer in bottles, forcing
Old Mike to end a long-standing routine and join Colonel Tom
with Coors Light in a can. I, however, finally caved and switched
to Michelob Light, the brand Donnery's mom, Dorothy, favored, a choice that seemed to earn fewer groans if not exactly
respect. "Even I used to drink it meself," Jim said. "At least it's
got a taste, luv." And Patty Hearst bought a brick mansion near
Guinan's, which got people wondering if maybe the media
baroness would come take on Fitz at the bar.

Oh, and Zip the rat finally passed on.

Meantime, the Station Plaza board didn't put John's name on
the lease right away. But they didn't ask him to leave either. One
thing everyone seemed to agree on was that so long as Jim was
alive, a certain peace would hold. And while the diabetes would
continue to curse his feet and later his eyes, the original barman
was feeling good enough to take a trip home alone to Ireland.

So John kept up his routine. He had no guarantees about his
future there, but he worked without them . . . did what he had,
or maybe more and more now, actually wanted to do. As the
weeks passed, he even started talking about getting around to
that business plan. And while his wife, Mary Jane, still woke up
most mornings with the other side of the bed empty, she found a
certain peace with her husband's choices. John made an effort to
get home earlier on the weekends and a couple of Sundays, Mary
Jane came to him, bringing with her geraniums to plant around
the store's patio.

Margaret was right by her brother's side, of course, changing their father's bandages, fixing his dinner, cleaning his laundry and keeping the inventory orders straight. In her mind, she might not have wanted her parents' life at the store, but in her heart she was bound there like metal to magnet and would never be able to let go completely. In that sense Guinan's continued to have a pop and a mom.

The human duct tape, meanwhile, proved strong in the years to follow with a revolving cast of characters who came and went through the old store's door. Andy helped hold down the fort for a while, struggling to broaden our musical horizons with operatic snippets from Verdi. When he left, a wry fellow named Bill arrived on the scene after his son, a commuter, suggested he might get a kick out of the place. Colonel Tom bartended sometimes; Ed helped paint the store and the Yannitellis's teenage son Anthony even voluntarily cleaned on the occasional Irish Night.

But the Guinans, they were our constant. Kelly, John's daughter, sometimes put in eight-hour days before taking a train to the city for her nighttime graphic design classes; even after she got a full-time job elsewhere in town, she kept working at the chapel on weekends and some evenings, protecting it the way her father and aunt had for so long. After his graduation, her brother Casey also became a regular fixture at the store. Because of the whole family, the 509ers still had somewhere to get a cup of coffee. Irish Night's musicians still had a stage. And Friday night's congregation continued plopping its souls down on the cracking green stools. Maybe nothing lasts forever, but this whole beautiful warm time that Colonel Tom had talked about, it lasts a while longer. And that's about as happy an ending as any of us could have hoped for.

ED AND I are at the bar with Fitz and the Count. Andy is serving beers in his Castrol Oil shirt and playing music from the band Morphine. Jim pokes his head in at one point to say hello, hears the funky slide bass beat and, shaking his head, heads back to the living room to turn up the TV and let Walker's heroism drown us out.

Kathryn's train is due any minute on the other side of the tracks. Getting up to leave, I notice Fitz's shirt collar is turned up. Without thinking, I reach over and pull it down. You look like the Fonz from Happy Days *like that, I tell him.*

"The Fonz," he repeats. "Heh, heh, heh, heh. *Hey, you know who else does that with my collar?"*

Who? I say.

"My daughter. She does that." He stands up and kisses me on the forehead.

As I'm pulling away in the car, I see Lou-Lou waddling toward the store's front door and dial Guinan's on my cell phone.

"Hello, Guinan's," Andy answers.

"Andy, Lou-Lou's at the door," I tell him, making the sharp right across the railroad tracks.

"What?" he says, turning down the music.

"Lou-Lou's at the door," I repeat, pulling into the parking lot and getting out.

"Oh, right," he says, understanding immediately. "Thanks."

Standing up on the platform, I hear stinging along the tracks announcing the train's arrival. And just before the engine noses around the bend, I see the little chapel's door opening—just wide enough—for her most faithful four-legged parishioner to squeeze inside.

acknowledgments

Much appreciation and love go to Joe Dizney and Jessie Woeltz for introducing Kathryn and me to Philipstown and to Jessie specifically for insisting we have that first beer at Guinan's. . . . To Linda and Lawton Johnson for the incalculable generosity of letting two strangers move into their home and lives after September 11. . . . To Sally and Bill Beatty, Nancy Cobb and Geoffrey Drummond, Gary Foster, Mindy Kramer and Anne Zehren for their friendship as well as the loan of clothes, cars and spare beds in the days following the attacks. . . . To Alison Berkley, Matt Johnson and Erle Norton for being among early readers of the manuscript. . . . To Cynthia Crossen and Jim Gleick for Sunday nights and Jim's unflappable technical assistance at all hours. . . . And to Walter and Jos Johnson for teaching me the intricacies of weeding, painting, and being a good neighbor and friend.

Many people helped with research of this book and loan of materials. I'd like to single out the staff at the Desmond-Fish Library, David and Cathy Lilburne of Antipodeon Books on Garrison's Landing, Chip Marks, Frederick Osborn III, Nora Preusser, her son Ed, Clemson Smith Muñiz, and Mary Ellen Yannitelli. Zenia Mucha worked wonders helping cut through the red tape around the governor. Thanks also to Dr. John Buse, director of the Diabetes Care Center at the University of North Carolina School of Medicine, for his guidance on the disease. And for anyone wishing to know more about the fantastic river and region around Guinan's, Frances F. Dunwell's book *The Hudson River Highlands* and Bill Moyers's documentary *America's First River* are two rich resources I used.

At the *Wall Street Journal,* I am indebted to Paul Steiger, who forgave my incoherent, scotch-laced pitch for a leave of absence and granted me time off to hang out in a bar. Thank you also to Melinda Beck for graciously supporting this project early on and to Larry Rout for making it possible for me to continue writing about the little chapels everywhere.

David Black is not only a tremendous agent, he is an astute reader and true lover of words—this book would never have evolved beyond the seed of an idea without him and his top-notch staff. At William Morrow, my editor, Rob McMahon, deserves credit not only for his sound guidance on the manuscript but also for his sound choice of drinking holes: he'd already been to Guinan's long before we ever met. Thank you also to Lisa Gallagher, Michael Morrison, Brian McSharry, and Rachel Bressler for supporting this project with enthusiasm, to Richard Aquan for art-directing such a beautiful cover, to Debbie Stier, Trina Rice, Samantha Hagerbaumer, and Heather Gould for steering me through publicity, to Erin Richnow for her steady hand, and to Trish Grader for her early belief in this story.

In my own family, I would like to note that my parents, Michael and Norva Bounds, have always known what matters most. I am grateful for their example, love and support of this book in many ways. Much love to my aunt, Catherine E. McKnight, and my grandmother, Catherine H. McKnight, who both knew and loved the original Pettiford, as well as to the other half of my gene pool: my grandfather Lee Bounds, who later in life got his own fish camp that brought us all together, my grandmother, Barbara Bounds Milone, my aunt, Bobbi Bounds Embree, and their spouses and partners in life, Jeff Surles, Edith Hubbard, Chuck Milone and Ken Embree.

To Kathryn Kranhold, thank you for making room for this new place and stage in our time together, for all the train rides to and from Garrison, and for the sound advice you have offered at the most critical junctures. You helped make my life in these years complete.

Finally and most important, my thanks to everyone who appears in this book for finding room at the bar for one more stranger and for loving this place enough to gallantly share your personal tales. My greatest gratitude goes, of course, to the Guinans themselves for all their candor and patient hours of interviews. They took me in as they have so many others since they arrived in America. Most people are lucky enough if they have one great family in their lives. I feel fortunate now to have two.

sources

Little Chapel on the River is based on my experiences during the course of several years at a country store and pub in Garrison, New York, called Guinan's. During that time, I spent hundreds of hours in formal interviews and informal conversations with members of the family who ran the establishment as well as past and current patrons of their establishment. I took notes conspicuously throughout my time there and kept a journal of my experiences. This is a work of nonfiction. Aside from a character called "the Sarge," no names have been changed and the dialogue is real and reconstructed with as much accuracy as possible given the sometimes rambling tempo of conversation inside a bar. The people featured prominently in these pages knew that I was writing a book in which they might be included; the major characters have all been privy to the passages where they are mentioned prior to the book's publication. However, any mistakes are mine alone.

I also relied on other printed and online sources for my research. The following is a truncated list of books, newspaper articles, websites and other materials that were important in developing the narrative. They offer a starting place for anyone wanting to delve more deeply into the rich legacy of the Hudson Highlands and places like Guinan's that give our towns grace and spirit.

Books

Dunwell, Frances F. *The Hudson River Highlands.* New York: Columbia University Press, 1991.

Erickson, Margery O. *A Few Citizens of Philipstown.* Garrison, N.Y.: Capriole Press, 1990.

—————. *Margaret S. Osborn: A Memoir.* Garrison, N.Y.: Capriole Press, 1992.

Horgan, John, and Rev. Frank Geer. *Where Was God on September 11?* San Francisco: BrownTrout Publishers, 2002.

Howell, William Thompson. *The Hudson Highlands*. New York: Walking News, 1982.

Lipsky, David. *Absolutely American*. New York: Houghton Mifflin, 2003.

Martin, James Kirby. *Benedict Arnold: Revolutionary Hero, an American Warrior Reconsidered*. New York: New York University Press, 1997.

Oldenburg, Ray. *Celebrating the Third Place: Inspiring Stories About the Great Good Places at the Heart of Our Communities*. New York: Marlowe & Company, 2001.

———. *The Great Good Place*. New York: Marlowe & Company, 1999.

Osborn, Frederick. *The Human Condition: How Did We Get Here and Where Are We Going*. Garrison, N.Y.: n.p., 1973.

Wood, Gordon S. *The American Revolution: A History*. New York: Modern Library/Random House, 2002.

Other Selected Works

Aig, Marlene. "GM Announcement Stuns Metropolitan Area Plant." Associated Press, February 24, 1992.

Barron, James. "Power Surge Blacks Out Northeast, Hitting Cities in Canada and Eight States; Midday Shutdowns Disrupt Millions." *New York Times,* August 15, 2003.

———. "Storm in the Northeast: The Overview; After a Day of Powdery Play, the City Faces Slushy Reality." *New York Times,* February 19, 2003.

Bounds, Wendy, and Kathryn Kranhold. "Amid the Ashes, Baby Carriages, Shoes, Family Photos: Displaced Apartment Dwellers Feel Fortunate to Be Alive, but Home Is Now War Zone." *Wall Street Journal,* September 14, 2001.

Braun, Richard. "Garrison Says Hello to 'Dolly' Film." *New York Sunday News,* June 16, 1968.

Curran, John. "Stranded Gamblers Suffer Blizzard of Losses." Associated Press, February 20, 2003.

Donohue, Pete, and Tracy Connor. "The Subways: A Special Hell." *New York Daily News,* August 15, 2003.

Garrison Naming. Putnam County Historical Society.

Gerwig, Henrietta, "Historic Philipstown in the Highlands of the Hudson." Putnam County Historical Society.

"Gloomy Return: U.S. Stocks Plummet as Trading Resumes Without Major Hitches—Investors, Unable to Shake Worries About Economy, Send Dow Down 7.13%—Sustaining Record Volume." *Wall Street Journal,* September 18, 2001.

"GM Closes Minivan Plant in North Tarrytown." *Buffalo News,* June 27, 1996.

Hernandez, Daisy. "With Plumbers' Candles and Guest Traffic Cops, Region Perseveres." *New York Times,* August 15, 2003.

"Hudson River and the Hudson River Railroad." Boston: Bradbury & Guild; New York: W.C. Locke & Co., 1851. (Reprint: Astoria, Queens, J.C. & A.L. Fawcett Inc.)

Industrial Bulletin. New York State Department of Labor, September 1968.

Ingrassia, Paul, and Joseph B. White. "GM Posts Record '91 Loss of $4.45 Billion and Identifies a Dozen Plants for Closing—Company Also Announces Its Second Restructuring in the Past Eight Years." *Wall Street Journal,* February 25, 1992.

Kahn, E. J., Jr. "The Hudson River." *Holiday,* October 1966.

Kilgannon, Corey. "A City's Jitters, Muffled in the Swirling Drifts." *New York Times,* February 18, 2003.

McFadden, Robert D. "Blizzard Buries Northeastern U.S., Disrupting Travel." *New York Times,* February 18, 2003.

Morfit, Cameron. "Weekender—Garrison, N.Y." *New York Times,* October 11, 2002.

Moyers, Bill. *America's First River: Bill Moyers on the Hudson. Part 1: Stories from the Hudson.* Public Affairs Television, April 23, 2002.

————. *America's First River: Bill Moyers on the Hudson. Part 2: The Fight to Save the River.* Public Affairs Television, April 24, 2002.

Osborn, Margaret. "Garrison's Landing Speech." Putnam County Historical Society, January 24, 1968.

Prevost, Lisa. "Creative People, at Home on Dirt Roads." *New York Times,* August 10, 2003.

Saunders, Jean. "Garrison's Landing." 1966.

"Streisand's Dolly: The $20 Million Film That Cannot Be Released." *Life,* February 14, 1969.

Winkler, Matthew. "Many Economists See Growth Crawling, but No Recession, After Stocks' Crash." *Wall Street Journal,* November 2, 1987.

Websites

"A Brief History of the Academy." www.usma.edu.

"The Great Chain." www.hhr.highlands.com.

"Hudson Estuary Basics." www.dec.state.ny.us.

"New York–New Jersey Harbor Estuary Program." www.harborestuary.org.

"Extraordinary Estuary." www.nywea.org.

"Natural History of the Hudson River: The River That Flows Both Ways." www.hhr.highlands.com.

"Henry Hudson." www.mariner.org.

"Who Was Henry Hudson Anyway? And What Happened to Him?" www.hudsonriver.com.

"A Curriculum of United States Labor History for Teachers." Sponsored by the Illinois Labor History Society. www.kentlaw.edu/ilhs.

"Railroad and Local History." www.hvrt.org/railhistory.

"Frederick Henry Osborn Papers." American Philosophical Society, Philadelphia. www.amphilsoc.org/library.

"About Putnam County." www.putnamcountyny.com.

"Early Putnam County History." www.hopefarm.com.

Made in United States
North Haven, CT
20 July 2023

39308285R00182